【国家智库报告】
Report of the National Think Tank

Plastic Pollution Prevention
and Control in China
Principles and Practice

中国塑料污染治理
理念与实践

张德元 等著

中国财经出版传媒集团

经济科学出版社
Economic Science Press

图书在版编目（CIP）数据

中国塑料污染治理理念与实践 / 张德元等著. —北京：经济科学出版社，2022.3
ISBN 978-7-5218-3537-3

Ⅰ.①中… Ⅱ.①张… Ⅲ.①塑料垃圾-污染防治-研究-中国 Ⅳ.①X5

中国版本图书馆CIP数据核字（2022）第049359号

责任编辑：孙丽丽　纪小小
责任校对：郑淑艳　徐　昕
版式设计：陈宇琰
责任印制：范　艳

中国塑料污染治理理念与实践
张德元 等著

经济科学出版社出版、发行　新华书店经销
社址：北京市海淀区阜成路甲28号　邮编：100142
总编部电话：010-88191217　发行部电话：010-88191522
网址：www.esp.com.cn
电子邮箱：esp@esp.com.cn
天猫网店：经济科学出版社旗舰店
网址：http://jjkxcbs.tmall.com
北京时捷印刷有限公司印装
710×1000毫米　16开　10.5印张　180000字
2022年4月第1版　2022年4月第1次印刷
ISBN 978-7-5218-3537-3　定价：58.00元
（图书出现印装问题，本社负责调换。电话：010-88191510）
（版权所有　侵权必究　打击盗版　举报热线：010-88191661
QQ：2242791300　营销中心电话：010-88191537
电子邮箱：dbts@esp.com.cn）

前言

塑料自从发明以来，凭借其优异的性能和低廉的成本，被广泛应用于人类生产生活当中，给人们的日常生活带来极大便利。但与此同时，塑料污染问题却日益严峻，特别是海洋塑料污染问题日益成为世界各国普遍关注的焦点环境问题，给人类的可持续发展带来极大挑战。联合国环境规划署发布的报告显示，海洋中的塑料废弃物总量为7500万~19900万吨，每年还以900万~1400万吨的数量在持续增加，如果不采取有效行动，到2040年，排放到水生态系统的塑料废弃物将增加近3倍。[1]

塑料和钢铁、橡胶等其他工业材料一样，是维持人类生存发展的一种重要材料，塑料本身并不是污染物，但如果废弃后的塑料制品管理处置不善，导致其泄漏到自然环境中并达到一定的水平，就会危害人类和自然界中动植物的生存和发展，成为严重的社会与环境问题。因此，塑料污染的本质是塑料废弃物管理不善造成的环境泄漏。

在可预见的未来，塑料仍将长期使用和存在。我们必须高度重视塑料污染问题，并加快探索塑料使用和生态环境相协调的可持续发展道路。诚如习近平总书记所言，当前国际社会日益成为一个你中有我、我中有你的"命运共同体"，面对世界经济的复杂形势和全球性问题，任何国家都不可能独善其身。当前，塑料污染与气候变化、生物多样性等生态环境问题共同成为全球面临的严峻挑战。

中国作为世界上最大的发展中国家，高度重视塑料污染治理，采取了一系列行动，开展了大量工作。经过不断探索和努力，逐渐形成了符合中国国情的有中国特色的塑料污染治理体系，即发展塑料循环经济，开展全链条治理：在设计生产环节，不断提升塑料性能、开发替代材料、开展产品生态设计、清洁生产；在

---

[1] *From Pollution to Solution: A Global Assessment of Marine Litter and Plastic Pollution*[R]. United Nations Environment Programme, 2021.

流通消费环节，出台了一系列促进绿色消费的政策措施，推广使用可循环使用产品和替代产品，减少一次性塑料制品消费；在消费后的处置利用环节，强化塑料再生利用和能源化利用，加强塑料废弃物规范处置，守住塑料环境泄漏的安全线。

经过几十年的努力，中国塑料污染治理取得了显著成效：一是建立起了由市场自发形成的覆盖广泛的废塑料循环利用体系，回收废塑料超过全球同期回收总量的45%[1]；二是依托完善的塑料工业体系，形成了覆盖高中低端的完善的再生塑料利用体系，材料化利用率达到31%[2]；三是中国不仅实现了本国塑料废弃物100%本土利用，在1992~2018年间还累计处置了来自欧盟、美国等其他国家和地区的1.06亿吨塑料废弃物[3]，并将其转化为再生塑料原料，为全球塑料污染治理做出了巨大贡献。通过发展塑料循环经济，中国为全球塑料的合理使用和人类的可持续发展探索出了一条可行的道路。

本报告的编写，旨在总结中国塑料污染治理的理念、实践和经验，为世界其他国家和地区的塑料污染治理提供借鉴，为全球塑料污染治理贡献中国智慧和中国力量。

本报告以中英文双语形式推出。在中文版的基础上，英文版进行了适当缩减。

声明： 本报告中的结论和观点，仅代表编写组意见，不代表任何政府部门的意见和建议。

---

[1] 根据世界主要国家公开数据及中国海关统计数据测算所得。
[2] 中国再生塑料行业发展报告2020-2021[R]. 中国物资再生协会再生塑料分会，2021.
[3] 依据中华人民共和国海关总署进口统计数据整理、计算所得。

Foreword

Since its invention, plastic has been widely used in production and daily life for its excellent performance and low cost, bringing great convenience to human beings. However, at the same time, the problem of plastic pollution is becoming increasingly serious, especially marine plastic pollution is becoming a focus of environmental concerns around the world, posing great challenges to the sustainable development of mankind. According to a report released by the United Nations Environment Programme, the amount of plastics in the oceans has been estimated to be around 75-199 million tons, with a 9-14 million tons increase year by year. And in the absence of necessary interventions, the amount of plastic waste entering aquatic ecosystems could nearly triple by 2040[1].

Plastic, similar to other industrial materials such as steel and rubber, is an important material for sustaining human survival and development, and the plastic itself is not a pollutant. However, if plastic products are not well managed and well disposed of after using and leaking into the natural environment and accumulating to a certain level, they may pose danger to human beings, plants, and animals' survival and development, therefore becoming a serious social and environmental issue. The essence of plastic pollution is its leakage into the natural environment caused by the mismanagement of plastic waste.

In the foreseeable future, plastics will still be widely applied for a long time. We need to pay great attention to plastic pollution and find out a sustainable development path that plastic use and the ecological environment could be balanced. As General Secretary Xi Jinping said, the international community is increasingly becoming a common destiny for all mankind. In the face of the complex world economy and global issues, no country can stand alone. At present, plastic pollution and climate change, biodiversity and other ecological and environmental issues are the severe challenge to all mankind.

As the world's largest developing country, China attaches great importance to plastic pollution control and has taken several actions. After consistent efforts, China has gradually formed the plastic pollution management system featured with Chinese characteristics based on its national conditions: conducting the whole-chain governance through plastic recycling economy, which means in the design and production process, it is required to constantly

---

1. *From Pollution to Solution: A Global Assessment of Marine Litter and Plastic Pollution*[R]. United Nations Environment Programme, 2021.

improve the performance of plastic, develop alternative materials, product eco-design, clean production; in the circulation and consumption process, a series of policies and measures are launched to promote green consumption. Recyclable products and alternative products are encouraged to reduce the consumption of disposable plastic products; in the after-use disposal process, it is proposed to enhance plastic recycling and energy utilization, strengthen the standardized disposal of plastic waste, and guard the bottom safety line of plastic leakage.

After decades of efforts, China has achieved remarkable results in plastic pollution control: Firstly, a recycling system covering a wide range of waste plastics has been established by the market voluntarily. The recycled waste plastics exceeded 45% of the total global recycling in the same period. Secondly, China has built a complete recycled plastics utilization system covering high, medium, and low ends based on its complete plastics industry system, of which the material and utilization rate is 31%. Thirdly, China has not only completely recycled the plastic waste produced domestically, but also disposed and recycled over 106 million tons of plastic waste from countries and regions such as the EU and the U.S. during 1992-2018, transforming them into recycled plastic materials and making a significant contribution to the global plastic pollution control. With the plastics recycling economy, China has shown the world a feasible way for recycling plastics and maintaining the sustainable development of human beings.

The report aims to provide China's reference in plastic pollution control for other countries through summarizing the concept, practice, and experience of plastic pollution control in China, and make China's contribution to global plastic pollution control.

Disclaimer: *The views and opinions expressed in this report are those of the authors and do not necessarily reflect the official policy or position of any agency of the Chinese government.*

专家学者寄语
Messages from Experts

　　随着社会、产业和科技的发展进步，人类在享受塑料功能性和经济性方面的红利很多年之后，现在到了强调更高标准的健康安全性和循环性的阶段。塑料污染治理，中国的行动意义重大。本报告对中国塑料污染治理的理念和实践所做的介绍和总结，既全面系统，又非常及时。相信中国的经验对于全球共同应对塑料污染这一世界性难题有非常重要的参考价值。

Human beings have been enjoying the functionality and economic benefits of plastics for years as society and technology have progressed. Now we've reached a stage where higher standards of health safety and recyclability are emphasized. China's actions are significant in the treatment of plastic pollution. This report presents a holistic and timely summary on the philosophy and practice of plastic pollution control in China. It is believed that China's experience can serve as a very important reference for the global treatment of plastic pollution.

——**金涌** Yong Jin

中国工程院院士、清华大学教授

Academician of Chinese Academy of Engineering, Professor of Tsinghua University

　　塑料已成为人类生产生活不可或缺的一类重要材料，但如何有效防止和减少全生命周期的环境影响，探索一条塑料的可持续发展道路，考验着人类智慧。中国在塑料污染治理上开展的努力、形成的经验十分宝贵，相信一定会为全世界塑料污染治理提供有益的借鉴。

Plastics has become an indispensable and important material in production and daily life, but how to practically reduce or prevent the negative impact of its entire life cycle and explore a sustainable development still remains a question. China has accumulated valuable experience in the treatment of plastic pollution, and I believe it will definitely provide a useful reference for the management of plastic pollution around the world.

——**左铁镛** Tieyong Zuo

中国工程院院士

Academician of Chinese Academy of Engineering

塑料的诞生，虽给人类生产生活带来了极大便利，但也给人类赖以生存的生态环境带来一定的危害，需要我们不断探索发展与保护的平衡，实现绿色可持续发展。塑料污染治理更需要树立绿色循环理念，从传统的生产环节延伸到流通、消费、回收利用、末端处置等全过程。在生态文明建设的背景下，加强国家统筹推进力度并引导社会力量广泛参与、加强国际合作并形成全球倡议来共同推动塑料污染治理，都是非常有意义的探索和实践。

The invention of plastics brings great convenience to human production and daily life, but also does harm to the ecological environment on which human beings depend for survival. We need to constantly maintain the balance between development and protection to achieve green and sustainable development. Plastic pollution treatment calls for the green circulation concept, extending from the traditional production process to the whole process of circulation, consumption, recycling, end disposal, and more. In the context of the construction of ecological civilization, it is an important exploration to strengthen national efforts and guide social forces to participate widely, strengthen international cooperation, and form global initiatives to jointly promote plastic pollution treatment.

——**徐光** Guang Xu
中华环境保护基金会理事长
Chairman of China Environmental Protection Foundation

世界经济论坛欢迎中国智库单位提出的解决塑料污染最佳实践的报告。与世界经济论坛促进循环经济和应对塑料污染挑战的使命相一致，本报告对控制塑料使用的实践过程提供了切实可行的见解，这些实践促进了利用生命周期方法来发挥塑料的效益，同时也保护我们共同的环境。这些工作方法代表了我们与中国之间进一步合作的机会，我们期待在本报告以及世界经济论坛系列伙伴关系的最佳实践的基础上进一步参与。

The World Economic Forum welcomes the initiative of the leading Chinese think tanks to address best practices of plastic pollution. Aligned with the mission of the WEF to promote circular economy and address the challenges of plastic pollution, this report provides tangible and practical insights to plastic control practices that promote life cycle approaches to leveraging the benefits of plastic while protecting our common environment. These working methods represent opportunities for further collaboration in China, and we look forward to further engagement building on this report and best practices developed across a range of WEF partnerships.

—— Gim Huay Neo
世界经济论坛董事会成员
Managing Board Member World Economic Forum

作者简介
Introduction to Authors

张德元 Deyuan Zhang

国家发展和改革委员会宏观经济研究院经济体制与管理研究所副研究员，长期从事生产者责任延伸制度、塑料污染治理、循环经济等相关领域的理论与政策研究。

Associate Researcher, Institute of Economic System and Management, Academy of Macroeconomic Research of National Development and Reform Commission. He has long been engaged in theoretical and policy research in related fields such as extended producer responsibility, plastic pollution control, circular economy, and so on.

王永刚 Yonggang Wang

中国物资再生协会再生塑料分会秘书长，长期从事废塑料回收再生行业政策与实践研究。

Secretary-General of Recycled Plastics Association, China National Resources Recycling Association. He has been long engaged in the policy and practice research of the plastics recycling industry.

彭绪庶 Xushu Peng

中国社会科学院数量经济与技术经济研究所研究员，中国社会科学院大学教授，主要从事产业技术创新与创新政策、绿色发展与生态文明、金融科技与信息服务等领域的研究。

Peng Xushu is a researcher at the Institute of Quantitative Technological Economics (IQTE), CASS and a professor at the University of the Chinese Academy of Social Sciences, mainly engaged in industrial-technological innovation and innovation policy, green development and ecological civilization, financial science, and technology and information services.

满娟 Juan Man

中国石油和化学工业联合会国际部高级工程师，国际交流与外企委员会副秘书长，长期从事石化行业国际交流与可持续发展研究。

Senior engineer of the International Department of China Petroleum and Chemical Industry Federation (CPCIF), Deputy Secretary-General of the Committee on International Exchange and Foreign Enterprises of CPCIF, has long been engaged in the study of international exchange and sustainable development of the petrochemical industry.

宋国轩 Guoxuan Song

国家发展和改革委员会国际合作中心重大活动处负责人，长期从事重大国际合作机制推进和落实、政府间官员交流培训等工作。

Head of the Major Event Division of the International Cooperation Center of the National Development and Reform Commission. He has long been engaged in the promotion and implementation of major international cooperation mechanisms and the exchange and training of intergovernmental officials.

袁嘉琪 Jiaqi Yuan

北京交通大学经济管理学院产业经济学博士研究生。

Ph. D. candidate, School of Economics and Management, Beijing Jiaotong University.

崔璇 Xuan Cui

北京交通大学经济管理学院硕士研究生。

Graduate student of School of Economics and Management, Beijing Jiaotong University.

范心雨 Xinyu Fan

华南理工大学工商管理学院技术经济及管理专业硕士研究生。

Graduate student of School of Business Administration, South China University of Technology.

专家顾问
Consultants

银温泉 Wenquan Yin

国家发展和改革委员会宏观经济研究院经济体制与管理研究所所长、研究员。

Director/researcher, Institute of Economic System and Management, Academy of Macroeconomic Research of National Development and Reform Commission.

李雪松 Xuesong Li

中国社会科学院数量经济与技术经济研究所所长、研究员，中国社会科学院宏观经济研究智库主任，中国社会科学院大学教授。

Director/researcher, Institute of Quantitative Technological Economics (IQTE), CASS; Director of Macroeconomic Research Think Tank, Chinese Academy of Social Sciences; Professor of University of Chinese Academy of Social Sciences.

李寿生 Shousheng Li

中国石油和化学工业联合会会长。

Chairman of China Petroleum and Chemical Industry Federation.

许军祥 Junxiang Xu

中国物资再生协会会长。

Chairman of China National Resources Recycling Association.

朱兵 Bing Zhu

清华大学循环经济研究院院长、教授，联合国环境署国际资源专家委员会委员，国家循环经济专家咨询委员会委员。

Dean/Professor of Institute of Circular Economy, Tsinghua University; Member of International Resource Panel of the United Nations Environment Programme; Member of the National Advisory Committee of Experts on Circular Economy.

吴玉锋 Yufeng Wu

北京工业大学教授，国家循环经济专家咨询委员会委员。

Professor of Beijing University of Technology; Member of the National Expert Advisory Committee on Circular Economy.

目录

01 1. 加强塑料污染治理的紧迫性

- 02　1.1 全球塑料污染问题日益严峻
- 05　1.2 中国同样面临着塑料污染问题

07 2. 塑料污染的本质和表现

- 08　2.1 塑料污染的本质
- 10　2.2 塑料污染的表现
- 15　2.3 塑料污染产生的原因

19 3. 中国塑料污染治理理念和路径

- 20　3.1 中国塑料污染治理的历史沿革
- 23　3.2 中国塑料污染治理的总体理念
- 25　3.3 中国发展塑料循环经济的基本路径

43 4. 中国塑料污染治理体系与成效

- 44　4.1 中国塑料污染治理体系
- 48　4.2 中国塑料污染治理取得的成效

57   5. 中国塑料污染治理实践的启示和借鉴

- *58*   5.1 应对塑料污染需要建立完善的全生命周期管理体系
- *59*   5.2 发展塑料循环经济需要建立完善基于本国的回收利用体系
- *61*   5.3 发展塑料循环经济需要综合考虑环境效益与经济效益
- *62*   5.4 发展塑料循环经济需要科学比较分析各种塑料替代产品和方案
- *63*   5.5 发展塑料循环经济需要加强政企合作和引导全社会广泛参与
- *64*   5.6 塑料污染治理需要不断加强国际合作

67   6. 加强全球塑料污染治理的倡议

71   结束语

- *72*   参考文献
- *74*   附录
- *82*   重点支持单位和企业
- *82*   致谢

Table of Contents

85 **1.**
Urgency of Strengthening Plastic Pollution Control

86 1.1 Global Pollution Issues Arising from Plastics are Increasingly Severe
88 1.2 China is also Faced with the Plastic Pollution Concern

91 **2.**
The Nature and Manifestations of Plastic Pollution

92 2.1 The Nature of Plastic Pollution
94 2.2 The Manifestations of Plastic Pollution
99 2.3 Causes of Plastic Pollution

103 **3.**
The Concept and Path of Plastic Pollution Control in China

104 3.1 Historical Evolution of Plastic Pollution Control in China
106 3.2 Overall Concept of Plastic Pollution Control in China
108 3.3 The Fundamental Path of Developing a Circular Economy for Plastics in China

113 **4.**
Plastic Pollution Control System and Accomplishments in China

114 4.1 Plastic Pollution Control System in China
117 4.2 Accomplishments of Plastic Pollution Control in China

123　5. Experience from Plastic Pollution Control in China

124 5.1 It is Necessary to Establish a Sound Life Cycle Management System to Control Plastic Pollution

125 5.2 To Develop Plastic Recycling Economy Calls for Home-based Recycling System

126 5.3 The Development of Circular Economy for Plastic Requires Comprehensive Consideration of Environmental and Economic Benefits

128 5.4 Developing Circular Economy for Plastic Requires Scientific Comparative Analysis of Various Plastic Substitute Products and Schemes

129 5.5 The Development of a Circular Economy for Plastics Calls for Cooperation between Government and Enterprises and a Holistic Participation

130 5.6 The Plastic Pollution Control Calls for Extensive International Cooperation

133　6. Initiatives on Enhancing Global Control of Plastic Pollution Control

137　Conclusion

138 References
140 Appendix
150 Institutes and Enterprises Who Offered Great Support to This Rreport
150 Acknowledgements

1.

加强塑料污染治理的紧迫性

塑料的发明是现代工业文明的标志之一,给人类的生产生活带来极大便利。但与此同时,塑料污染问题也日益严峻,成为世界各国普遍关注的焦点环境问题。

1.1 全球塑料污染问题日益严峻

塑料自诞生以来,便凭借其质量轻、制造成本低、可塑性强等特点,广泛应用于工业、农业、服务业等国民经济各个领域,融入人类生产生活的方方面面,其"身影"无处不在,与钢铁、木材和水泥等一起成为现代经济社会发展的基础材料。

根据欧洲塑料制造商协会(Plastics Europe)的分析数据[1],2015~2020年,全球塑料产量和消费量以每年平均2%的速度稳定增长,产量从2015年的3.2亿吨增长到2020年的3.67亿吨;人均消费量从2015年的43.63千克增长到2020年的46.60千克。预计到2035年,塑料产量将增加1倍,到2050年产量将增加2倍,届时,全球年人均塑料消费量将达到84.37千克(见图1-1)。

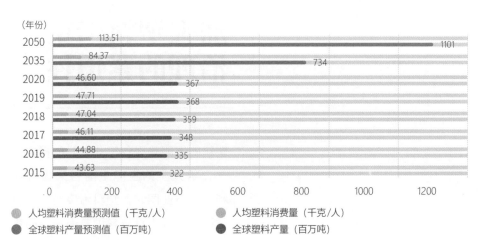

图1-1 全球塑料产量和消费量
资料来源:欧洲塑料制造商协会。

---

[1] 欧洲塑料制造商协会, https://plasticseurope.org/knowledge-hub/.

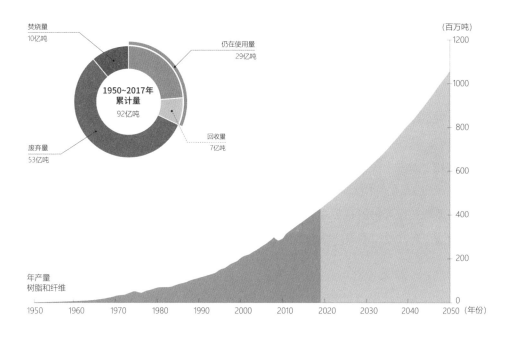

图 1-2 全球塑料累计量和处置方式构成

资料来源：*From Pollution to Solution: Global Assessment of Marine Litter and Plastic Pollution*[R]. United Nations Environment Programme,2021.

随着塑料生产量、消费量的快速增长，塑料的废弃量也在快速增加。对此，各国际组织开展了大量的分析和评估。联合国环境规划署2018年发布的报告显示，全世界塑料废弃物年产生量约为3亿吨[1]，大量的塑料废弃物进入土壤和海洋，最终形成"白色污染"。如图1-2所示，1950~2017年期间全球累计生产约92亿吨塑料，预计到2050年，全球塑料累计产量将增长到340亿吨，年均增长率达到7.9%。[2]

---

[1] United Nations (UN) Environment，*Beat Plastic Pollution*[EB/OL]. 2018, https://www.unenvironment.org/interactive/beat-plastic-pollution/.

[2] *From Pollution to Solution: Global Assessment of Marine Litter and Plastic Pollution*[R]. United Nations Environment Programme,2021.

塑料从材料本身看，可回收性非常高，但由于塑料包装等塑料制品质量轻、价值低，对其回收一般没有经济性，因此全球实际上能够被回收的塑料占比很低，其余大部分被堆积在垃圾填埋场或直接遗弃在环境中。这些泄漏到环境中的塑料废弃物，自然分解和矿化过程十分缓慢，会在环境中不断累积，造成"不可逆"的生态环境破坏，产生严重的塑料污染问题。

当前，从沙漠到农田，从高山之巅到海底深处，甚至在遥远的南北极，塑料废弃物几乎无处不在。根据经济合作与发展组织（OECD）的统计数据，2019年全球约产生3.5亿吨塑料废弃物。[1] 尽管全球大部分国家采取了积极有效的应对措施，但塑料废弃物的排放量仍然逐年持续增长。[2] 如此严重的塑料污染问题在全球普遍存在，亟须采取广泛而有效的措施加以应对。

---

[1] OECD统计数据，https://stats.oecd.org/Index.aspx#.

[2] Lau W. W. Y. et al., Evaluating scenarios toward zero plastic pollution[J]. *Science*, 2020,369:1455-1461.

1.2 中国同样面临着塑料污染问题

塑料污染是全球面临的共同挑战，也是各国的共同责任，尽管中国作为发展中国家，人均塑料消费量远低于发达国家和地区，但同样面临着塑料污染的挑战。

1.2.1 中国塑料消费量逐年增长

随着中国经济发展和生活消费水平的提高，塑料的消费量持续增长。另外，随着近年来人们消费方式的变化和新兴领域的快速发展，以及新冠肺炎疫情导致对一次性医疗及防护用品需求的大量增加，一次性塑料制品使用量也快速增长。2020年，中国塑料用量为9087.7万吨，同比增长12.2%。[1]

中国作为全球第一大商品出口国，在为全球提供优质商品的同时，这些塑料包装等塑料制品也随着各类商品出口到世界各地，为其他国家和地区人民的生产生活提供便利，成为全球经济社会发展与后疫情时代绿色复苏的重要物质保障。

1.2.2 中国塑料污染治理压力日益增大

随着中国经济的发展和生活消费水平的提高，塑料的消费量持续增长。另外，随着近年来人们消费方式的变化，电商、快递、外卖等新兴领域的快速发展，以及新冠肺炎疫情导致对一次性医疗及防护用品需求的大量增加，一次性塑料制品使用量也快速增长。随着塑料消费量的不断攀升，塑料废弃物逐渐增多，由此带来的环

---

[1] 中国再生塑料行业发展报告2020-2021[R].中国物资再生协会再生塑料分会，2021.

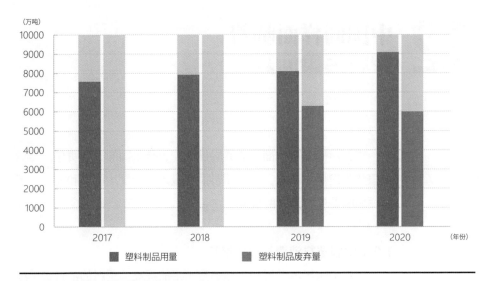

图 1-3 2017~2020 年中国塑料制品用量及废弃量
资料来源：根据《中国再生塑料行业发展报告2020-2021》和马占峰等发表在《中国塑料》期刊上的年度文章《中国塑料加工工业2020》整理所得。

境污染压力也逐渐增大。如图1-3所示，近年来，中国塑料废弃物年产生量超过6000万吨，其中除材料化利用和能源化利用外，剩余的基本还是通过与其他生活垃圾一并填埋的方式处置，不仅占用了大量的土地，而且随着时间的推移，塑料废弃物有泄漏的风险，给生态环境安全带来一定的威胁。

当前，塑料污染问题继气候变化之后成为全球关注的又一焦点环境问题。中国正在加快推动生态文明建设，致力于探索工业文明后人类经济社会发展与生态环境相协调的生态文明发展道路。中国愿意也理应与世界各国一道，探索全球塑料污染治理的有效路径。

2.

塑料污染的
本质和表现

塑料废弃物一旦泄漏到自然环境中，在自然状态下，需要数百年甚至上千年才能完全降解，将对全球土壤环境、水生态环境、气候变化、生物多样性等带来长期不利影响。

2.1 塑料污染的本质

塑料是当代社会重要的基础材料，给人民群众的日常生活带来了极大方便，广泛应用于工业、农业、服务业等诸多领域。塑料本身并不是污染物。塑料污染的本质是塑料废弃物不当管理造成的环境泄漏。当前，塑料污染问题如此严重是历史长期累积的结果。

2.1.1 塑料污染的本质是塑料废弃物的环境泄漏

2019年，全球塑料的体积相当于钢铁和水泥总和的26%，且消耗量以每年2%的平均速度增长[1]，可以说塑料已经成为当今社会不可或缺的重要原料。塑料与钢铁、有色金属等其他工业材料一样，其本身具有很好的可回收性，理论上完全可以进行回收和再生循环利用，从而成为一种重要的再利用资源，避免泄漏到自然环境中。

但是，塑料应用领域非常广泛，类型复杂多样，其中一些塑料制品在消费后很容易被丢弃，难以收集，存在泄漏到水体、土壤等自然环境中的风险。如超薄农用地膜在使用后"一扯就碎"，回收难度大，易造成土壤污染。再如餐饮单位堂食服务中使用的一次性塑料刀、叉、勺和吸管，容易混入餐厨垃圾，给餐厨垃圾厌氧发酵等后续资源化处置利用带来困难。再如添加塑料微珠的淋洗类化妆品，其中的塑料微珠会随着污水系统进入自然环境。再加上塑料本身具有很好的耐腐蚀性，即使是小小的塑料吸管，也会在自然条件下长期存在，长年累积就会形成较严重的污染。相较于其他材料带来的环境问题，治理塑料污染更为复杂。

---

[1] 郑强.塑料与"白色污染"刍议[EB/OL].https://mp.weixin.qq.com/s/qLPyAVI2xt49-QiUCOHdPA,2021/11/19.

2.1.2 全球塑料污染形成具有长期的历史累积性

自20世纪六七十年代以来，塑料产业快速发展，全球塑料生产量、消费量增长迅速。在这一阶段，可持续发展理念尚未达成全球共识，各个国家和地区尚未普遍建立起针对塑料废弃物的政策法规和出台相应的管控措施，缺乏回收和再生利用的技术手段，同时在产品生产端也未贯彻生态设计理念。多种因素共同导致了在很长的历史时期内，塑料废弃物因管理不当产生了大量的环境泄漏，造成污染。因此，塑料污染之所以如此严重，是长期历史累积的结果。

随着各国立法工作的推进，产品生态设计的广泛推行，废弃物管理体系的完善，回收基础设施的建设，回收和资源化利用技术的创新，我们相信人类有能力对塑料污染进行有效管控。

2.2 塑料污染的表现

塑料一旦泄漏到土壤、水体等自然环境中，便难以降解，会造成视觉污染、土壤污染、水体污染等各种环境破坏（见图2-1），给脆弱的生态环境带来持久性危害。

2.2.1 塑料废弃物带来的视觉污染

"视觉污染"最初用于形容城市中刺眼的广告牌对风景的破坏，后广泛用于描述城市中各种杂乱的环境造成的视觉不良观感。塑料的视觉污染是指散落在环境中的塑料废弃物对城市容貌和自然景观造成的破坏。[1] 塑料袋、塑料膜、废农膜等一些塑料制品和包装物比较轻薄，容易被风吹起，漂浮在空中或挂在树枝上，出现在一些旅游景区和道路两侧，形成"白色污染"。当前，在一些欠发达国家和地区，大量的塑料废弃物仍都没有得到有效收集和妥善处置，散落在自然环境中，严重危害生态环境安全，成为自然环境中的不和谐"音符"。

需要指出的是，处于"视觉污染"阶段的塑料废弃物还没有达到真正"不可逆"的程度，若人们能及时对散落在自然环境中的塑料废弃物进行有效收集和合理处置，这些"白色污染"就会被扼杀在"视觉污染"阶段，而不会进一步在自然界中崩裂分解，形成微塑料[2]，最终泄漏到自然水体和土壤当中，给生态环境造成更深层次的危害。

---

[1] 韩立钊,王同林,姚燕."白色污染"的污染现状及防治对策研究[J].中国人口·资源与环境, 2010, 20(S1): 402-404.
[2] 微塑料：直径小于5毫米的塑料碎片与颗粒。

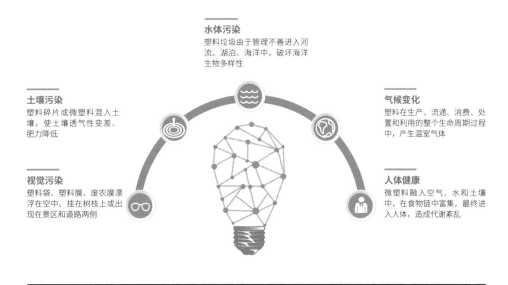

图 2-1 塑料污染的表现

2.2.2 塑料废弃物带来的水体污染

水体塑料污染是指塑料废弃物由于管理不善通过各种途径进入河流、湖泊、海洋当中，并影响水生态环境的现象。塑料水体污染主要分为陆源塑料水体污染和海源塑料水体污染，其中陆源塑料水体污染主要是由于塑料废弃物直接进入各类水体中，或者以微塑料的形态进入水体中带来的污染；海源塑料水体污染主要来源于渔业和海产养殖过程中的各种渔具、航运和海上作业使用的海洋设备和生产生活垃圾，以及游客携带的各类塑料用品在使用后丢弃到海洋当中形成的污染。水体中的塑料污染要比土壤中的塑料污染更加隐蔽，其影响也更为广泛（见图2-2）。

这些流入海洋的塑料废弃物，经过永不停息的海洋运动以及风化作用，会分解变成微塑料参与到物质循环中，抑或悬浮在水中，被海洋生物误食而进入生物圈，又抑或沉入海底，等待漫长的自然分解过程，会对海洋生物多样性造成极大破坏。有"海洋中的热带雨林"之称的珊瑚礁，是世界上独一无二的生态系统，90%以上的海洋生物都直接或间接地依赖于此，而如今珊瑚礁正面临海洋塑料废弃物的威胁。科学家经过研究发现，珊瑚礁一旦与塑料进行接触，其患病风险从4%增加到89%，受到疾病侵袭的可能性将增加20倍。除了珊瑚礁以外，其他海洋生

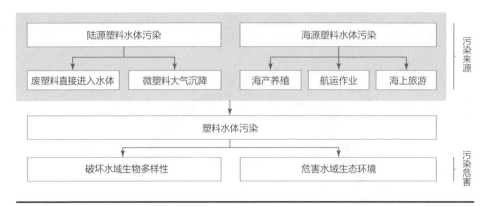

图 2-2 水体塑料污染路径

物也面临同样的"塑料噩梦"。例如，处于食物链底部的浮游生物，一旦误食微塑料，将会阻碍自身正常进食，干扰机体正常反应，存活率和繁殖率大大降低；一些较大型的海洋生物，很容易被"幽灵渔具"缠住，导致身体受损而引发感染和死亡。不仅如此，由于这些塑料废弃物表面附着了大量的微生物和藻类，释放出一种令海洋生物食欲大开的味道，同时其颜色和形状与水母等海洋生物类似，导致海洋生物吞食塑料废弃物的情况时常发生。据马修·麦克劳德等（Matthew Macleod et al.）2021年发表在 *Science* 杂志上的文章显示，由于塑料废弃物缠绕和吞食塑料废弃物，国际自然保护联盟红色名单上693个物种中的约118个受到严重威胁。[1]

2.2.3 塑料废弃物带来的土壤污染

相较于"视觉污染"，土壤当中的塑料污染则隐蔽得多。土壤塑料污染是指塑料以塑料碎片或微塑料的形态混入土壤当中，致使土壤改变原来的性能和状态，如土壤透气性变差、肥力降低等。土壤塑料污染主要来自塑料废弃物、道路径流中包括的轮胎磨损颗粒以及农用地膜和农药瓶等农用物资在使用后的随意丢弃。此外，使用含微塑料的畜禽粪污堆肥、使用含有微塑料的污泥作为肥料、使用含微塑料的污水用于灌溉等也会将微塑料带入土壤当中（见图2-3）。

---

[1] Matthew MacLeod, Hans Peter H. Arp, Mine B. Tekman, Annika Jahnke. The global threat from plastic pollution [J]. *Science*, 2021: 6550.

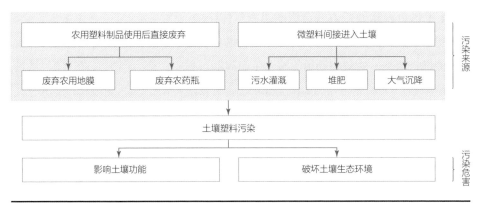

图 2-3 土壤塑料污染路径

据马修·麦克劳德等 2021年发表在 *Science* 杂志上的文章显示,目前土壤中的塑料含量可能比海洋表面的塑料含量还要高,其塑料组分占土壤有机碳的比例已经高达0.1%。[1] 一项对污水处理厂的调查发现,约90%的微塑料在污水处理后会累积到污泥中。在北美和欧洲地区,约50%的污泥用于农业生产,据此估算,北美地区每年通过污泥农用化而进入土壤中的微塑料量为6.3万~43万吨,欧洲地区为4.4万~30万吨。[2]

土壤中不断积累的塑料废弃物不仅会影响土壤透气性,造成土壤板结,阻碍植物根系生长,而且会逐渐分解成微塑料,影响土壤中微生物的代谢活动。更严重的是,微塑料进入土壤后,在长期风化作用、紫外光照射以及土壤中其他组分的共同作用下,会大量吸附土壤中的重金属和杀虫剂、除草剂、抗生素等有机污染物,并沉积在土壤环境中,损害土壤生态系统的健康。与此同时,土壤生物的活动会加速微塑料的二次分解和迁移扩散,形成微塑料在土壤中传播的恶性循环。[3]

---

[1] Matthew MacLeod, Hans Peter H. Arp, Mine B. Tekman, Annika Jahnke. The global threat from plastic pollution [J]. *Science*, 2021: 6550.

[2] Nizzetto L., Futter M., Langaas S.. Are agricultural soils dumps for microplastics of urban origin?[J]. *Environmental Science & Technology*, 2016, 50: 10777-10779.

[3] Horton A. A., Walton A., Spurgeon D. J., et al.. Microplastics in freshwater and terrestrial environments: Evaluating the current understanding to identify the knowledge gaps and future research priorities [J]. *Science of the Total Environment*, 2017, 586: 127-141.

2.2.4 塑料废弃物处置方式会影响温室气体排放

塑料制品本质上是由化学物质和化石燃料转化而来的，这就意味着塑料在生产、流通、消费、处置和利用的整个生命周期过程中，都会产生温室气体。从具体数据来看，全世界每年有2.8亿~3.6亿吨化石燃料用于生产塑料[1]，在这一过程中，化石燃料中的碳元素便转移到了塑料中。如果能对这些塑料废弃物进行有效回收和再生循环利用，就可以避免其中所含的碳释放到自然环境中，从而减少温室气体的排放。但如果这些塑料废弃物被降解、焚烧或填埋，其中的剩余碳元素最终将以二氧化碳、甲烷等温室气体的形态逐渐释放出来，给温室气体减排造成压力。

2.2.5 微塑料可能会对人体健康带来危害

塑料污染不仅给我们的生态环境造成了严重破坏，也可能会威胁到人类自身的健康。塑料废弃后，会分解成肉眼看不到的微塑料，融入空气、水和土壤当中。这些泄漏到环境中的微塑料极易被植物吸收，被鱼类和小动物误食，从而进入食物链中，通过食物链进行传递并在各级食物链逐级富集，直到进入人类体内（见图2-4）。此外，塑料微珠在牙膏、沐浴露等日化洗护用品中使用广泛，随着塑料制品的使用，一些微塑料会被人体直接摄入。甚至有研究认为，摄入的微塑料将进入人体的循环系统并到达特定的组织，有可能会造成氧化应激、炎症反应和代谢紊乱等，甚至影响DNA信息的表达和遗传。[2]

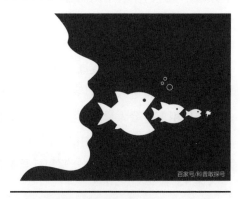

图2-4 微塑料累积传导示意图
资料来源：https://baijiahao.baidu.com/s?id=1691641723892993408&wfr=spider&for=pc。

---

[1] Dees J. P., Ateia M., Sanchez D. L.. Microplastics and their degradation products in surface waters: A missing piece of the global carbon cycle puzzle [J]. *ACS ES&T Water*, 2020(1): 214-216.

[2] Landrigan, P.J., Stegeman, J., Fleming, L., Allemand, D., Anderson, D., Backer, L. et al.. Human health and ocean pollution[J]. *Annals of Global Health*, 2020(1): 1-64.

2.3 塑料污染产生的原因

在我们的日常生产生活中，人工合成材料种类繁多、使用广泛，但为什么塑料污染问题会成为全球普遍关注的焦点环境问题呢？造成这一现象的原因需要从塑料本身的属性和塑料不当处置两方面进行具体分析。

2.3.1 塑料材料的难分解属性

塑料在自然条件下很难分解，导致塑料污染具有累积性。大部分塑料都属于短消费周期产品，特别是一次性塑料制品，消费使用后很快被丢弃。但塑料制品物理化学结构稳定，在自然环境中完全被微生物同化，降解成二氧化碳（CO_2）和水，实现无机矿化可能需要200~400年的时间。根据有关研究结果，聚苯乙烯这种塑料在土壤、污泥、腐烂垃圾或粪肥微生物群落里，4个月仅降解掉0.01%~3%。[1] 正是由于塑料自然降解过程漫长的特性，导致塑料废弃物一旦泄漏到环境中就会逐渐累积，日积月累使得"小塑料"变成了"大危害"。

2.3.2 塑料材料及制品的多样性

（1）塑料种类庞杂，给废弃后的分类回收和再生利用带来困难。

随着塑料工业的快速发展，塑料种类也变得越来越庞杂。我们常见的就有聚乙烯（PE）、聚丙烯（PP）、聚氯乙烯（PVC）、聚苯乙烯（PS）、丙烯腈—丁二烯—苯乙烯共聚合物（ABS）、聚

---

[1] Yu Y., Yang J., Wu W. M., et al.. Biodegradation and Mineralization of Polystyrene by Plastic-Eating Mealworms: Part 2. Role of Gut Microorganisms[J]. *Environmental Science & Technology*, 2015, 49(20): 12087-12093.

甲基丙烯酸酯（PMMA）、乙烯—醋酸乙烯酯共聚合物（EVA）、聚对苯二甲酸乙二醇酯（PET）、聚酰胺（PA）、聚碳酸酯（PC）、聚氨酯（PU）、聚甲醛（POM）、聚乳酸（PLA）、聚四氟乙烯（PTFE）等。除此之外，还有大量的多层塑料复合薄膜、纸塑复合薄膜、铝塑复合薄膜等复合材料的存在。这些塑料制品在废弃后需要进行分类回收和再生利用。庞杂的塑料品种给塑料废弃后的分类收集和再生利用带来极大挑战。

(2) **许多塑料制品体积小、质量轻、回收成本高、利用价值低。**

当前，塑料不仅用于生产飞机、轮船等大型产品，在包装等领域也广泛应用。一些小的食品包装仅有几克重，体积小、质量轻，对这些小型塑料制品进行回收需要付出大量的人工等经济成本，但获得的经济收益却非常少，导致对塑料包装和塑料棉签等小型塑料制品的回收利用不具备经济价值。即使考虑到对这些小型塑料制品回收利用的资源节约和环境保护效益，也很难形成有效的运行机制促进这一目标的实现。因此，大量的塑料包装在废弃后被焚烧或填埋，甚至随意丢弃到自然环境中，使大自然成为塑料污染的"重灾区"。

2.3.3 塑料制品废弃后的不当处置

由于塑料的自然分解需要漫长的时间，因而塑料污染形成过程更加隐蔽而缓慢。相较于其他固体污染物，塑料污染不像大气污染、水污染，在短时期内就会对生态环境带来较严重的影响，可以很容易被人们感受到，从而及时采取措施加以应对。

(1) **人类的不当行为造成的塑料环境泄漏。**

在我们的日常生活中，塑料的使用非常普遍。我们吃的药、食品是用塑料包装的，日常用的牙签、棉签等也是塑料制成的，农业生产中会使用大量的农膜、地膜、农药瓶，海洋作业中需要塑料制成的渔网、渔具等。这些日常生产生活中使用的塑料制品，容易被人为地随意丢弃。这些有意无意的塑料废弃物随意丢弃行为，给塑料废弃物的回收和处置带来极大困难，容易形成环境泄漏。可以说，我们每个人的不当行为都是导致塑料污染的直接原因。

(2) 塑料废弃物回收利用和处置设施建设不完善。

每一个国家都应当对自己国家产生的塑料废弃物进行合理收集和处置。但当前世界上大部分国家的塑料废弃物回收利用和处置设施建设都不完善。有些国家缺乏再生利用设施，塑料废弃物仍需出口到其他国家和地区进行回收利用，由于长途运输的成本增加，导致低价值的塑料废弃物不具备回收利用经济价值，从而整体材料化回收利用率较低；有些国家塑料废弃物焚烧设施、规范的填埋设施处置能力严重不足，导致大部分塑料废弃物仍以填埋为主，或排放到自然环境中。

这样的状况，在如今的欧盟、美国等发达国家和地区也一样严峻。2018年，欧盟28国以及挪威和瑞士共30国共产生2910万吨塑料废弃物，其中通过焚烧发电实现能源回收、通过再生利用实现材料化回收和通过直接填埋处置的重量分别为1240万吨、946万吨和725万吨，占比分别为42.6%、32.5%和24.9%。[1] 2018年，美国各类塑料废弃物产生量达到3568万吨，但只有309万吨被材料化回收，占比仅为8.66%；能源化回收量为562万吨，占比15.75%；填埋量高达2697万吨，占比75.59%。[2] 这些被填埋的塑料废弃物会随着时间的推移逐渐分解，很容易泄漏到环境当中，成为塑料污染的潜在来源。

(3) 塑料废弃物跨境转移带来的环境风险。

目前，面对日益严峻的塑料污染，一些国家并没有迅速采取有效应对措施，主动承担本应自己国家承担的塑料污染治理责任，仍有许多国家将本国收集的塑料废弃物出口到其他国家，给其他国家和地区塑料污染治理带来极大压力。据统计，2020年全球塑料废弃物出口总量高达385万吨，其中出口量最多的10个国家总共出口了260万吨，占出口总量的67.5%，美国是出口最多的国家，出口量为62万吨，荷兰紧随其后，为34.6万吨，这些塑料废弃物大部分出口到了土耳其、马来西亚等地，给当地的生态环境带来一定威胁。[3]

---

[1] *Plastics - the Facts 2020: An Analysis of European Plastics Production, Demand and Waste Data*[R]. Plastics Europe Association of Plastics Manufactures, 2020.

[2] U.S. Environmental Protection Agency, https://www.epa.gov/facts-and-figures-about-materials-waste-and-recycling/plastics-material-specific-data.

[3] UN Comtrade Database.

中国塑料污染治理理念和路径

中国作为世界上最大的发展中国家，高度重视塑料污染治理，通过大力发展塑料循环经济，开展塑料全链条治理，探索出了一条具有中国特色的塑料污染治理路径。

3.1 中国塑料污染治理的历史沿革

中国在古代就树立了"天人合一"的思想，崇尚自然、尊重自然，倡导人与自然和谐共生的理念。因此，在漫长的发展过程中，不断探索各种环境问题的有效解决路径。对于塑料污染问题，更是在很早就开始重视，并采取措施积极应对。

(1) 以重点领域治理为主的早期塑料污染治理阶段。

改革开放以后，随着中国经济的快速发展和人民生活水平的不断提高，塑料消费量快速增长，由此引发的"白色污染"问题逐渐显现。中国这一阶段的塑料污染治理主要突出问题导向，以对个别环境问题的治理为主，包含塑料包装废弃物、一次性发泡塑料餐具、塑料袋等。如1989年9月，原国家环保局、建设部、铁道部、交通部、国家旅游局联合印发了《关于加强重点交通干线、流域及旅游景区塑料包装废物管理的若干意见》，以重点领域塑料包装治理为主，禁止在铁路车站、旅客列车、客船和旅游船上使用一次性发泡塑料餐具，杜绝塑料包装废弃物在河流、湖泊中及沿岸堆积，同时各部门广泛宣传"白色污染"的危害性，向塑料污染治理"宣战"。再如，针对一次性发泡塑料餐盒使用量大、使用后收集处理难等突出问题，2001年4月，中国政府有关部门发布《关于立即停止生产一次性发泡塑料餐具的紧急通知》，禁止使用一次性发泡塑料餐具。2007年12月，中国政府发布了《国务院办公厅关于限制生产销售使用塑料购物袋的通知》，规定自2008年6月1日起，在所有超市、商场、集贸市场等商品零售场所实行塑料购物袋有偿使用制度，一律不得免费提供塑料购物袋。该措施有效缓解了当时严重的

"白色污染"问题，全国大型超市一年的塑料袋使用量减少了约2/3。[1]

(2) 以发展塑料循环经济为核心的全面治理阶段。

20世纪90年代末，中国逐渐步入工业化发展中期，资源环境对经济社会发展的约束日益突出，中国逐步引入循环经济理念，推动经济社会可持续发展。这一背景下，中国逐渐认识到塑料污染治理不能"头痛医头，脚痛医脚"，必须开展系统化治理，强化塑料废弃物回收和循环利用。2005年，中国政府有关部门开展循环经济示范试点建设，其中包括废塑料循环利用试点示范建设项目。2009年，中国正式实施了《循环经济促进法》，全面推动覆盖全社会的资源循环利用体系建设。在这一理念引领下，中国按照"3R"原则［即"减量化（reduce）、再利用（reuse）、资源化（recycle）"］开展塑料污染治理，并逐渐由重点领域治理向全面循环发展阶段过渡。

在源头减量方面，2009年1月，中国政府发布了《国务院办公厅关于治理商品过度包装工作的通知》，提出禁止生产、销售过度包装商品，并按照减量化、再利用、资源化的原则，从包装层数、包装用材、包装有效容积、包装成本比重、包装物回收利用等方面，对商品包装进行规范，引导企业在包装设计和生产环节中减少资源消耗，降低废弃物的产生，方便包装物回收和再利用。

在末端回收方面，2010年5月，中国政府有关部门发布了《关于开展"城市矿产"示范基地建设的通知》，推动包括塑料废弃物在内的报废机电设备、电线电缆、家电、汽车、手机、铅酸电池、塑料、橡胶等重点"城市矿产"资源的再生循环利用。2017年5月，农业部颁布《农膜回收行动方案》，加大对废弃农膜、地膜的治理，推动废农膜、废地膜的回收和再生利用。

(3) 开展塑料污染全链条治理的新发展阶段。

这一阶段，快递、外卖、电商等新兴领域快速发展，一次性塑料制品用量快速增加，塑料污染日益严峻。为有效应对塑料污染问题，中国全面加强塑料污染治理，开展全链条治理。2020年1月，国家发展和改革委员会（以下简称"发改委"）、生态环境部出台《关于进一步加强塑料污染治理的意见》，提出了覆盖塑料制品生产、消费、流通、回收利用等各个环节的治理政策措施，推动中

---

[1] 限塑一年全国超市塑料袋减少了2/3 [EB/OL]. 华西都市报, https://news.ifeng.com.mainland/special/2010lianghui/zuixin/201003/0311_9417_1571533_1.shtml, 2010/03/11.

国塑料污染治理进入全链条管理新阶段。2021年9月，发改委会同有关部门印发《"十四五"塑料污染治理行动方案》，从塑料生产和使用源头减量，科学稳妥推广塑料替代产品，加强塑料废弃物规范回收和清运，建立完善农村塑料废弃物收运处置体系，加大塑料废弃物再生利用力度，提升塑料废弃物无害化处置水平，加强江河湖海、旅游景区、农村地区塑料废弃物清理整治等方面，对2021~2025年的塑料污染治理重点工作进行了全面部署。

2020年修订的《中华人民共和国固体废物污染环境防治法》规定：电子商务、快递、外卖等行业应当优先采用可重复使用、易回收利用的包装物，优化物品包装，减少包装物的使用，并积极回收利用包装物；国家鼓励和引导减少使用、积极回收塑料袋等一次性塑料制品，推广应用可循环、易回收、可降解的替代产品；旅游、住宿等行业应当按照国家有关规定推行不主动提供一次性用品等；产生秸秆、废弃农用薄膜、农药包装废弃物等农业固体废物的单位和其他生产经营者，应当采取回收利用和其他防止污染环境的措施。

在多年的塑料污染治理实践中，中国塑料污染治理的法律和政策体系不断完善，覆盖的领域和范围不断拓展，治理力度也越来越大，逐步形成了全链条的闭环管理体系（见图3-1）。

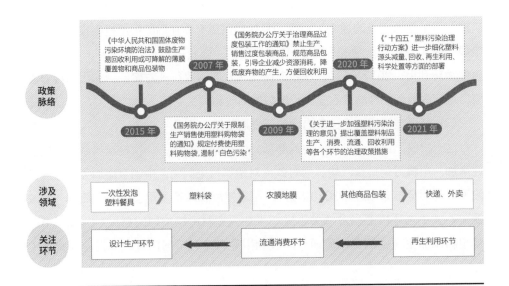

图 3-1 中国塑料污染治理历史脉络

3.2 中国塑料污染治理的总体理念

作为世界上最大的发展中国家，中国在几十年高速发展过程中面临的资源环境约束更加突出，走可持续发展道路的决心更加坚定。在这样的背景下，中国抛弃传统的以资源能源高消耗、污染物高排放为特征的"大量生产、大量消费、大量废弃"的线性经济增长模式，大力发展循环经济，探索经济增长与资源环境相脱钩的可持续发展模式。

中国是世界上第三个颁布循环经济专门法律的国家，与德国循环经济聚焦于固体废弃物管理不同，中国通过在生产、流通、消费各环节发展循环经济，在生产、流通、消费各环节和全过程贯彻落实"3R"原则，逐步构建起"资源—产品—再生资源"的循环经济发展模式，大幅降低了原生资源使用量，实现了资源利用的最大化和对生态环境负面影响的最小化。

塑料是人类生产生活中必不可少的材料，中国要实现工业化、建设社会主义现代化强国，将塑料从我们的生产生活中去除既是不科学的，也是不现实的。因此，塑料的使用和消费必然会导致废弃物的产生。面对潜在的塑料污染风险，必须加强塑料废弃物的回收和利用，发展塑料循环经济，加快构建从塑料设计生产到废弃后回收利用和处置的闭合式循环发展模式，探索塑料使用与生态环境保护的协调发展之路。

按照循环经济"3R"原则，首先是"减量化"，尽可能减少一次性塑料制品的生产和使用，特别是一些末端无法回收处理的塑料制品更要严格限制使用，从源头上大力推动塑料制品易回收、易再生等生态设计方法；其次是"再利用"，在流通消费环节，探索可循环的塑料制品及商业模式，如推广可循环托盘、加大可循环包装

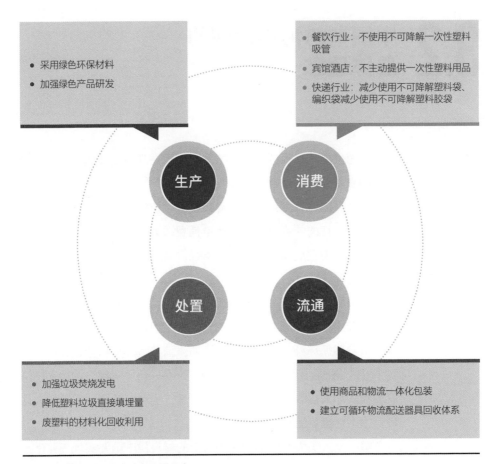

图 3-2 中国塑料污染全链条治理理念

物的使用比例等；最后是"资源化"，在末端处置环节，开展废弃塑料制品的回收和材料化利用，对暂时不具备材料化利用条件的进行能源化回收利用，最大限度地减少塑料废弃物的填埋量，进行"无害化"处置，从而构建起覆盖塑料污染全链条的治理体系（见图3-2）。

3.3 中国发展塑料循环经济的基本路径

中国塑料循环经济的发展，经历了从短期抑制到长期循环体系构建，从强调末端处置到全链条治理的逐步完善的过程。在几十年的发展过程中，逐步探索出具有中国特色的塑料污染全生命周期治理体系和治理路径，强调在塑料原料生产、塑料制品生态设计、塑料制品消费、塑料废弃物回收利用、最终塑料废弃物安全处置等全过程进行有效治理和不断改进（见图3-3）。

图 3-3 中国发展塑料循环经济基本路径

3.3.1 改进和创新塑料原料

在原材料生产阶段,通过对塑料材料的改进和创新,针对不同的应用场景,提升各种原材料的性能,开发更加绿色环保的原材料,可以从源头预防和减少可能产生的塑料污染。

(1) 减少材料中有毒有害物质添加。

要避免塑料成为污染物或者降低废弃后带来的污染,就要减少材料生产环节的有毒有害物质添加。一方面,可以通过研发更高性能的树脂材料,减少其中残

通过提升材料性能实现"源头减塑"

塑料材料开发环节的绿色发展是塑料循环经济发展的重要组成部分。某全球性的材料科技企业,近年来提出诸多有助于推动可持续发展的材料和包装解决方案,并推动这些方案在下游产业中的使用,从而助力行业"绿色变革"。

【延长材料使用寿命】

• ENGAGE™ PV聚烯烃弹性体(POE)薄膜,可以使太阳能组件的使用寿命从25年延长到30年,提高光伏发电装置发电量、可靠性和使用寿命,提高光伏组件抗电势诱导衰减性能,降低度电成本以及总系统成本。

• 通过在配方中加入AFFINITY™聚烯烃弹性体(POE),应用于可重复使用的塑料托盘,可以使使用寿命延长1.5倍,相较于传统木制托盘或者注塑托盘,其全生命周期可以降低1821.779千克二氧化碳当量排放,减排效应明显。

【助力包装材质单一化】

该企业提出的INNATE™ TF双向拉伸聚乙烯树脂(TF-BOPE)和全PE结构包装解决方案,用来替代传统的多种材质的复合包装生产,目前已经应用于某企业产品线生产。据预测,如果洗涤行业都使用单一化学结构的塑料软包装,可助力实现其回收和再生利用,每年可减少5万~6万吨的塑料用量。

留的有毒单体和裂解物，降低树脂本身的毒性；另一方面，在材料生产中减少有毒有害溶剂添加，如能增强可塑性的增塑剂，抑制材料发生氧化反应的抗氧化剂，降低摩擦系数或减少静电干扰的抗黏连剂，以及着色剂、稳定剂、爽滑剂等。

(2) 提升材料性能以提高可回收性。

塑料制品使用环境复杂，特别是一些农业、渔业使用制品，使用后容易破碎损坏，不易收集。通过提升塑料性能可以有效提高塑料制品废弃后的可回收性。具体可以通过在塑料中添加合适的改性剂，提高塑料产品使用寿命，提升材料性能，减轻环境压力。如通过提升材料性能减少多层复合包装物的生产和使用；通过提升农用地膜的韧性来提高废旧地膜的可回收性，从而减少地膜残留对土壤的危害等。

(3) 探索开发新型替代环保材料。

科学合理地开展一次性塑料制品绿色替代产品开发，可以有效减少对塑料制品的消耗。目前，市场上出现了许多竹木制品、纸制品等环保包装和可重复使用的购物袋等，用以替代一次性塑料制品。也有企业正在研发生物可降解塑料以替代传统塑料，用于吸管、购物袋的生产和使用。

3.3.2 推行塑料制品生态设计

研究表明，在设计阶段，充分考虑现有技术条件、原材料保障等因素，优化解决各个环节资源环境问题，可以最大限度实现资源节约，从源头减少环境污染。[1] 因此，在塑料制品设计环节，融入全生命周期管理的理念，对塑料制品在生产、流通、消费和废弃后的回收利用或处置环节的社会、经济与环境影响开展系统评价，并采取措施改进产品设计，以达到尽可能减少塑料制品在全生命周期过程中的环境影响，实现资源节约和环境改善的目的。

---

[1] 工业和信息化部、发展和改革委员会、环境保护部.关于开展工业产品生态设计的指导意见[R].工业部节能与综合利用司，2013.

加大再生材料使用

某国内最早践行ESG[Environmental（环境）、Social（社会）、Governance（公司治理）]的企业，在产品生态设计领域积极推进产品绿色低碳设计方案创新，协同供应链及行业力量，为社会绿色发展贡献力量。

【在产品中广泛应用再生塑料】

克服信息通信技术（Information and Communication Technology，ICT）产品设计挑战，实现了在笔记本、台式机、工作站、显示器及外设产品中广泛使用再生塑料。自2005年起，企业产品中共使用了毛重超过1.15亿千克的再生塑料。

【开展闭环再生塑料应用实践】

提出产品回收处置与再生材料应用的闭环管理理念。通过产品设计创新、供应链协同合作、材料溯源管理等措施实现电子废弃物中废塑料的闭合循环。2017年，在台式机和显示器产品上成功使用闭环再生塑料。2020年，将闭环再生塑料的使用范围扩大至103种产品。目前，累计使用了接近900万千克来自电子废弃物中的再生塑料材料。

使用全生命周期评估方法评估发现，相比于原生材料，通过使用闭环再生材料：（1）逐步切断了对化石原料的依赖，每生产1千克 JH960-6900 再生材料减少使用0.81千克原油；每生产1千克 GAR-011（L85）再生材料减少使用1.72千克原油。（2）有效减少了产品全生命周期中的二氧化碳排放，每使用1千克 JH960-6900材料的碳减排量为2.67千克，比原生材料降低45.33%；每使用1千克 GAR-011（L85）材料的碳减排量为2.923千克，比原生材料降低75.34%。

改进设计减少原材料使用

某企业自2017年以来，在全球启动了"一代人的可持续发展计划"。在塑料包装可持续利用方面,就源头减塑和重复使用包装创新开展了积极探索。

【源头减塑】

2020年,秉承"减少不必要包装"的原则,对口香糖的塑料包装展开"瘦身"计划。在保证产品包装安全性和稳定性的前提下,将口香糖包装瓶盖和瓶身承载物之间的空隙进行压缩,缩减包装的体积。仅这一项计划就帮助公司每年减少了450吨的原生塑料使用。2021年,进一步对20多个产品包装进行优化迭代,每年将减少580吨的原生塑料使用。

【重复使用包装】

2021年,联合线下零售商家,在上海、广州、武汉、宁波和南昌5个城市的17家门店和部分大型购物商场推出了可以重复使用的巧克力豆金属罐和纸盒包装。消费者在购买时,商家将提供可重复使用的包装盒,并告知消费者再次购买时,如携带原包装盒将享受优惠。

开展塑料制品"双易设计"

废塑料的再生循环利用依赖于良好的产品设计。目前,为了满足大众的消费需求和产品的标新立异,很多塑料制品的设计从材质、颜色、标签甚至到形状,都出现多样化、复杂化趋势,不利于废弃后回收,导致废塑料回收率普遍较低。同时,回收的废塑料再生利用一般都需要进行除杂、分选,塑料制品使用的材料、结构和标签信息等,对再生利用的影响至关重要。因此,对塑料制品进行"易回收、易再生"设计,可以从根本上提高废塑料的回收利用率。

中国石油和化学工业联合会与中国物资再生协会联合16家企业共同发起成立了"绿色再生塑料供应链联合工作组"(Green Recycling Plastic Supply Chain Group,GRPG),旨在推动塑料制品的"易回收、易再生"设计,并制定了《塑料制品易回收易再生设计评价通则》,从塑料制品的易回收、易再生性质入手，通过产品设计与回收表现之间的关联关系评价并指导塑料制品的设计，以达到"突破制约塑料回收率提高的'瓶颈'"的目的。

自2021年2月发布《塑料制品易回收易再生设计评价通则》以来，已经有立白、宝洁、超力包装、利德宝和广州瑞远等著名企业完成了包装物"双易认证"。

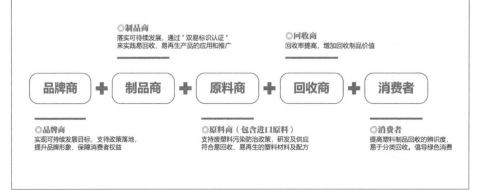

（1）材料的单一化设计。

由于塑料材质种类非常多，要实现其材料化回收利用还需要对各类塑料进行分类分选。相对于复合材料来说，采用单一材料生产的产品在废弃后不需要对复合材料进行分离分解，因此更容易对其进行回收和再生利用。如果塑料制品在设计生产过程中，尽量使用单一塑料材质来替代多塑料材质的原料选择，会显著提高塑料制品的可回收性，而且使用的塑料种类越少，产品的结构和生产工艺就越简单，这样不仅可以降低产品生产过程中的能耗以及污染物排放，还能使废弃后的塑料制品在回收利用过程中更容易被回收和再生利用。

（2）再生塑料的推广使用。

塑料的再生循环利用，不仅可以减少对原生资源的消耗，还可以大幅减少生产过程中的资源和能源消耗，有效减少塑料制品全生命周期的温室气体排放量，避免塑料制品废弃后的环境泄漏。因此，塑料制品在设计生产时更多地使用再生塑料作为原料，打通塑料废弃物再生循环的"产业通道"，可以有效构建起闭环的塑料循环利用体系，是推动末端塑料废弃物循环利用的关键。

(3) 重复使用的塑料制品设计。

塑料污染之所以这么严重，主要是因为一次性塑料制品占比大，这些产品往往使用一次后就丢弃，造成大量浪费。因此，在设计环节通过改进产品设计，更多地采用可重复使用的产品设计方案，减少一次性塑料产品占比，再配合某些商业模式变革，用更多的可循环使用塑料制品替代一次性塑料制品，可以有效减少塑料消耗和由此带来的污染。比如物流快递行业，通过增加其使用的中转箱和中转袋的强度实现循环使用，开展可循环快递包装规模化应用，推广标准化物流周转箱循环共用，可以大幅减少一次性塑料编织袋的使用。

(4) 易回收利用的塑料制品设计。

从材质本身看，塑料制品理论上都可以实现再生循环利用。但在实践中，由于末端回收利用环节分类分选成本高，再生循环利用的经济性差，从而导致其实际再生循环利用率低。如果塑料制品在设计时就充分考虑到废弃后的拆分方便，更多地使用简单易拆解、易分离的结构设计，可以有效降低末端回收利用的成本投入，便于末端的回收与再利用，从而达到节省回收成本、减少污染排放、提高回收率的目的。比如某些产品采用模块化结构设计，在废弃后很容易拆卸和分离，从而降低拆解处理成本；再如在饮料包装上，通过使用易剥离的黏合剂粘贴塑料瓶表面的商标，或通过易撕扯设计来提高其易回收性。

3.3.3 减少不必要的一次性塑料制品消费

社会公众作为塑料制品消费的主要群体，通过推行绿色消费模式，减少对一次性塑料制品的使用对塑料污染治理至关重要。

(1) 减少一次性塑料购物袋使用。

当前，在世界范围内一次性塑料购物袋的使用非常普遍，成为塑料污染治理中的"小物品、大麻烦"，回收利用难，环境泄漏风险大。对此，中国出台了一系列法律法规和规范性要求，要求一定区域的商场、超市、药店、书店、集贸市场等经营性服务场所，对一次性塑料购物袋进行收费，通过价格杠杆限制和调节一次性塑料购物袋的使用。另外，消费者也是一次性塑料袋的主要使用者，通过鼓励消费者使用环保布袋、纸袋等非塑料制品和可重复使用的购物袋，可以有效减少一次性塑料袋的使用和消费。

"青山计划"

随着中国外卖行业快速发展，网络订餐成为更多人的就餐方式，但与此同时所引发的环境问题备受关注。为加快推动外卖行业环保进程，2017年某外卖平台企业发起关注环境保护的"青山计划"行动，从一次性餐盒的使用、回收、再利用等多维度进行探索，打通从源头减量、废弃物回收、循环再生的产业链条，在非必要包装减量、绿色包装替换和餐盒再生循环利用方面开展积极探索。

【鼓励用户绿色消费】

率先上线"无须餐具"选项产品功能，推进实施下单"必选项"，配以多产品场景激励机制，如"配捐绿色能量""领取数字人民币红包"等，开展每月一天"外卖环保日"宣传，累计激励了超过1亿外卖用户选择"无须餐具"。

【引导商户绿色经营】

上线"商家青山档案"，逾200万商户向各方展示自身减塑环保努力。发布指南手册搭建商户绿色经营知识体系，在多省市开展"青山可持续餐饮示范街"建设。

【推动绿色包装解决方案】

积极探索符合餐饮商户需求的绿色包装方案，联合包装生产企业等开展外卖包装创新和孵化，累计向商户免费投放28款91万件创新环保包装制品。

【开展绿色包装供应链建设】

推出"青山计划绿色包装推荐名录"，纳入三大类101家生产企业的161个环保产品，搭建商家绿色包装采购平台，持续推动包装绿色化发展。

（2）减少一次性塑料餐饮具使用。

由于塑料成本低廉、性能优异，在餐饮领域，塑料吸管，塑料刀、叉、勺，塑料餐布，塑料碗，塑料杯，塑料碟等一次性塑料餐具的使用非常广泛。中国出台了一次性塑料限制使用的规范性文件，通过在餐饮行业减少使用一次性塑料吸管，在餐饮饭店提供堂食服务过程中，减少使用一次性塑料刀、叉、勺、水杯、碗碟等餐具，鼓励在确保卫生安全的前提下，提供可重复使用的餐饮器具，有效减少一次性塑料餐具的使用，从而避免其废弃后带来的环境污染。

（3）减少酒店一次性塑料制品使用。

当前，为追求生活便利，由塑料制成的一次性拖鞋、牙刷、梳子、洗发水瓶、沐浴液瓶等，在各个酒店广泛使用，造成巨大的资源浪费和环境危害。中国制定出台了相关法规和规范性文件，要求一定区域的宾馆、酒店等经营场所不得主动提供一次性洗护用品、免费的一次性拖鞋等一次性塑料用品；鼓励宾馆、酒店等住宿服务提供商，通过提供需要付费购买的便携式洗漱包、采用续充式洗护用品等方式提供相关服务；鼓励消费者自备洗护用品等，从而减少酒店住宿环节一次性塑料制品的消费。

（4）减少电商、快递、外卖等新兴领域塑料制品消费。

随着商业模式的变化，如今电商、快递、外卖等新兴行业蓬勃发展，一次性塑料包装的使用量也随之增加。中国针对电商、快递、外卖等新兴领域出台了专门规定，鼓励电子商务、外卖等平台企业和快递企业制定及实施一次性塑料制品减量计划，鼓励消费者点餐过程中减少餐具使用，发布绿色包装产品推荐目录，推进产品与快递包装一体化，推广电商快件原装直发，减少电商商品在寄递环节的二次包装，减少包装过程中一次性塑料封套的使用，推广使用"瘦身"胶带和可循环使用的包装物等，减少上述领域一次性塑料制品的消费。

塑料零部件	有价包装物	农药瓶和农膜	低值废塑料
建设完善的回收利用体系，加强其中塑料的分类回收，提高回收率	通过从事废弃塑料瓶回收的个人和企业，建立起覆盖城市、乡村的回收体系，实现应收尽收	实施生产者责任延伸制度，推动农业生产经营者履行农药包装和地膜回收处理义务——押金制。	将塑料垃圾与其他生活垃圾进行分类，从而得到材质统一、相对洁净的塑料垃圾

图 3-4 不同类型废塑料回收利用模式

3.3.4 对塑料制品进行科学的分类回收

分类回收是实现塑料循环利用的关键环节。在开展分类回收过程中，应当根据废弃物的产品特性、流通特点、处置去向、经济价值等，采用不同的回收模式（见图3-4）。

（1）开展塑料零部件回收。

随着各种改性塑料的不断出现，塑料以塑料零部件的形式在家电、汽车等产品中的应用越来越广泛，家电中塑料占比为23%~57%[1]，汽车中塑料占比约为6%[2]。这些产品废弃后价值较高，已经形成了完善的回收利用体系，其中的塑料零部件可以随着这些产品的回收而被很好地回收和再生循环利用。因此，对这一部分塑料制品，可以采取"伴随回收"的模式，建设完善的废弃电器电子产品、报废汽车等废旧产品回收利用体系，并加强对这些产品中塑料零部件的分类回收，从而有效提高这部分塑料制品的回收率。2020年，中国回收的电器电子产品和报废汽车中废塑料回收量达到了240万吨。[3]

---

[1][2][3] 中国再生塑料行业发展报告 2020-2021[R].中国物资再生协会再生塑料分会，2021.

(2) 开展有价包装物回收。

在中国，城市塑料废弃物中废弃塑料瓶的占比达到22%，其中以PET饮料瓶、HDPE日用产品包装桶（瓶）等为主。这些废弃塑料包装，连同一部分PP塑料餐盒，目前仍具有一定经济价值。为获取这部分经济价值，社会上存在大量的从事废弃塑料瓶和部分塑料餐盒回收的个人和企业，建立起了覆盖城市、乡村的庞大的回收利用网络，基本上能够实现废弃塑料瓶的"应收尽收"，并长期维持较高的回收率水平。因此，对这部分塑料制品应采取"专门回收"模式。目前，在饮料瓶回收领域，中国涌现出了智能回收机回收、"互联网+回收"等新型回收模式，对进一步提高回收效率发挥了重要作用。2020年，中国仅PET瓶的回收量就达到了380万吨。[1]

(3) 开展农药瓶和农膜回收。

塑料农药瓶和农膜使用范围广，收集难度大，环境泄漏风险高，是塑料污染治理的难点。对这部分塑料制品应采取"强制回收"的模式。2020年，中国制定实施了《农用薄膜管理办法》，规定农用薄膜使用者应当在使用期限到期前捡拾田间的非全生物降解农用薄膜，交至回收网点或回收工作者，不得随意弃置、掩埋或者焚烧；农用薄膜生产者、销售者、回收网点、废旧农用薄膜回收再利用企业或其他组织等应当开展合作，采取多种方式，建立健全农用薄膜回收利用体系，推动废旧农用薄膜回收。同年，制定实施了《农药包装废弃物回收处理管理办法》，规定农药生产者（含向中国出口农药的企业）、经营者和使用者应当积极履行农药包装废弃物回收处理义务，及时回收农药包装废弃物并进行处理。农药生产者、经营者应当按照"谁生产、经营，谁回收"的原则，履行相应的农药包装废弃物回收义务。目前，在中国的一些地方也出现了由第三方专业公司运营管理，采用押金制回收农药瓶的新模式。

(4) 做好低值废塑料的回收。

低值废塑料包括塑料包装袋、塑料包装膜、一部分塑料餐盒等，因其回收处理成本高、再生材料品质低等原因大多不具备经济价值。其中，塑料餐盒如果能够实现集中回收和有效分类，仍具备一定的再利用经济价值。因此，对这部分塑

---

[1] 中国再生塑料行业发展报告2020-2021[R].中国物资再生协会再生塑料分会，2021.

外卖餐盒回收再生示范项目

某领先的互联网外卖平台企业，不断探索外卖塑料餐盒回收利用途径。目前，联合相关企业和地方政府开展垃圾分类及餐盒回收试点，配套建设餐盒分拣线，促进餐盒高值化再生利用。目前已在多个省份建立常态化餐盒回收体系，实现塑料餐盒的规模化回收和"重塑新生"。

【多场景回收再生探索】

基于热点订单数据，在识别回收可行性较强的区域开展落地，先后在北京、上海、广州等多个省市开展1200余个垃圾分类及餐盒回收试点，覆盖城市、校园、社区、写字楼、餐饮门店、景区等7类场景。回收后的塑料餐盒被制作成单车挡泥板、日历壳、名片、钥匙扣等再次循环利用。

【城市级回收示范实践】

2021年3月于厦门启动首个城市级餐盒回收再生试点，年回收折合约30%的厦门外卖包装减塑量；该项目覆盖厦门4个辖区，8个街道，1000余社区及单位。

【本地智能化回收路径】

在上海示范项目中，开通餐盒回收预约功能，外卖用户可通过外卖App—垃圾分类—预约回收入口，将收集到的塑料餐盒下单预约上门回收服务。同时，布点社区的前端智能回收设备也增加餐盒回收品类，外卖用户可前往投递。

提升材料性能促进地膜回收利用

目前，地膜广泛应用于农作物生产，覆盖玉米、马铃薯、水稻、棉花、番茄、烟草等作物，对保障粮食安全、促进农民增收发挥了重要作用。但同时，由于地膜本身强度和耐老化性能差，作物收获后地膜破碎严重、可回收

> 性差，引起严重的塑料残留污染问题。
>
> 　　某企业通过提高原材料性能，生产高性能茂金属聚乙烯，在强度、韧性等性能方面显著改善，可以在不增加厚度的情况下，大幅提升地膜强度，并保持较好的光学和保温保墒性能。通过在新疆喀什地区30万亩棉田覆膜种植实验，该地膜在使用6个月后残膜回收率达90%以上，比传统地膜回收率提高30%。同时，通过产业链合作，实现了回收地膜的再资源化，将"白色塑料污染"变成"绿色再生资源"，从根本上解决了农业增产、农民增收和污染治理问题，为农业可持续发展做出了积极贡献。

料制品可以采取"两网融合、统一回收"的模式，通过将这些塑料废弃物与其他的生活垃圾进行分类，可以使得一些材质统一、相对洁净的塑料废弃物，如塑料餐盒等重新进行物理回收，从而提高整个塑料制品的再生利用率。另外，如果能对不同材质的塑料包装进行分类收集，也将有助于对其进行化学回收。

3.3.5 实现废塑料的再生循环利用

回收的各类废弃塑料制品，只有经过再生利用才能真正实现由"垃圾"到"资源"的华丽转身，实现"塑料原料—塑料制品—流通消费—再生塑料"的生命周期循环。

（1）废塑料的物理再生。

对废弃塑料制品进行物理再生，使废塑料变为再生塑料原料重新用于生产，是废弃塑料制品处置的最优选择。废塑料的物理再生（也叫物理循环或物理回收）指将预处理后的废塑料通过熔融造粒等物理方式加工为再生原料的方法。物理再生简单可行，一般分为熔融再生和改性再生。熔融再生主要用于回收成分单一、相对干净、易于清洗的废塑料，将废塑料经过分选、破碎、清洗、干燥、造粒得到再生塑料颗粒。改性再生是通过物理或化学改性的方法提高废塑料的抗冲击性、耐热性、抗老化性等，以及改善其颜色和外观，拓宽废塑料的应用渠道，提高其应用价值。另外，对于一些城市生活垃圾中的一次性薄膜等难分选废塑料制品，利用价值较低，且表面粘有较多油脂等有机质，清洗过程会产生大量污水

通过全产业链合作实现"塑尽其用"

贯彻绿色低碳循环发展理念,推动废塑料的回收和再生循环利用,发展塑料循环经济是实现塑料可持续发展的重要路径,也是塑料污染治理的优先选择方案。

【建立广泛的回收体系】

中国某专业从事废塑料回收利用的企业,与全球数百家废弃物收集商和拆解商紧密合作,参与回收来源于城市废弃物、工业废弃物、农业废弃物中的废塑料,小到废弃的饮料瓶,大到报废的冰箱、汽车等。经初步分选后,这些废塑料进入企业的处理工厂,经过精细化分选、多途径清洗、造粒、改性等工艺,让废塑料重获"新生",成为绿色新材料。

【不断拓展下游应用渠道】

目前,企业每年为全球提供包括再生PP、再生PE、再生ABS、再生PC、再生PS、再生PA、再生PBT、再生PET等在内的十余种环保高性能再生塑料粒子,已广泛应用于家用电器、包装(含日化、超商、快递、电商等)、家居、纺织、汽车、IT、电子、电动工具、建筑、电气、能源等行业。其"塑尽其用"方案,致力于通过循环回收体系实现塑料的再生利用,助力国内外品牌商落地其中长期可持续战略。

【取得显著资源环境效益】

通过"塑尽其用"的解决方案,不仅防止塑料废弃物流入环境中,而且也免于填埋和焚烧,实现了经济效益、环境效益和社会效益的协同。2004~2020年期间,该企业共生产超过120万吨再生塑料,相比原生塑料,累计约减少石油使用620万吨,减少煤炭使用168万吨,减少二氧化碳排放152万吨,减少用电56亿千瓦时,减少用水8000万吨。

且难以清洗干净，因此该类废塑料一般不直接用来造粒，而是通过与木粉等无机填料复合制备木塑复合材料。

(2) **废塑料的化学再生。**

除了可以进行物理回收的塑料废弃物外，还存在着大量的膜袋类塑料废弃物，这些塑料废弃物很难通过物理再生的方法进行回收。因此，塑料废弃物化学再生技术应运而生。塑料废弃物化学再生（也叫化学循环或化学回收）指利用化学技术使塑料废弃物重新转化为树脂单体、低聚物、裂解油或合成气的方法，可分为热裂解回收法和化学分解回收法。对废塑料进行热裂解处理能够实现资源的循环利用，有效提高塑料废弃物的处理效率，减少废塑料填埋和焚烧引起的环境污染。借助先进的技术装备开展化学再生可以生产出与原生塑料同品质的产品，是塑料废弃物实现高值化利用的路径之一。但目前，化学回收还没有成熟的案例，仍需要进一步开发能适应不同塑料原料、活性高、选择性好的成熟工艺技术，并探索建立与之相适应的回收体系和可持续商业模式。

3.3.6 开展塑料废弃物能源化利用

无法进行材料化回收利用的塑料废弃物会进入生活垃圾处理体系，这些混杂塑料难分选、难清洗，在现有技术经济条件下很难回收，只能采取能源化利用方式进行焚烧发电，获取其能源利用价值。废塑料具有较高的燃烧热值，如聚乙烯和聚丙烯的燃烧热值分别高达46.63GJ/kg和43.95GJ/kg，均高于木材的燃烧热值14.65GJ/kg，通过焚烧可大幅减少塑料废弃物的堆积量和填埋土地占用，使废塑料减容90%~95%。[1] 但同时，生活垃圾中的塑料废弃物还含有少量聚氯乙烯、聚丙烯腈、聚氨酯等，这些塑料在燃烧时会产生有害物质和温室气体。例如，PVC在燃烧时产生氯化氢气体，聚丙烯腈和聚氨酯在燃烧时会产生氰化氢。因此，在焚烧过程中做好污染物控制至关重要。

---

[1] 侯彩霞.超临界水降解塑料的研究[D].天津大学，2003.

3.3.7 加强消费行为引导与绿色教育

绿色消费是指各类消费主体在消费活动过程中贯彻绿色低碳理念的消费行为。塑料污染的治理同样离不开社会公众的广泛参与。中国在塑料污染治理过程中，始终强化对消费者行为的宣传引导和绿色教育。

(1) 加强塑料污染的宣传引导。

早在2007年，中国塑料污染治理政策发布后，社会各界就广泛加强宣传。有的地方组织开展了"袋袋相传"活动，倡导使用环保布袋，减少使用塑料制品，杜绝"白色污染"，树立环保、绿色生活理念。在2020年新版塑料污染治理政策发布之后，各地、各社会组织开展的绿色消费周、DIY环保袋、环保袋设计大赛等各类活动层出不穷，有的商场、超市在醒目位置张贴承诺书和宣传标语，显示自身"限塑"的决心，让顾客能了解有关政策，引起了社会各界广泛关注，使绿色环保理念得以传播，让"绿色、低碳、环保"的理念深入人心。

(2) 重视对中小学生的绿色教育。

中小学生是社会未来的消费主力，从小培养其绿色消费理念对推动全社会形成绿色消费新风尚具有重要带动作用。为了进一步培养中小学生塑料污染防治意识，有的小学开展了"倡导绿色生活，塑料污染治理从我做起"环保主题进校园宣讲活动，生动形象地讲述了塑料污染的由来，呼吁学生从点滴小事做起，并通过言行号召家人和朋友，做生态环保的宣传者和践行者，让绿色消费的理念"代代相传"。有的地方组织开展了"生态环境宣传进校园"系列活动，呼吁广大中小学生能够在日常生活当中做小小宣传员，通过自身力量，带动周边人一起减少塑料制品的使用，加强环保意识。

江西省南昌市人民政府驻厦门办事处关于"减少使用塑料制品"倡议书*：

（1）不使用一次性塑料饭盒、一次性餐桌布和一次性筷子。

（2）重拎布袋子、重提菜篮子，重复使用耐用型购物袋，拒绝使用一次性塑料袋。

（3）购物时自备购物袋，尽量减少使用一次性纸杯、木筷和餐盒，以及一切可以产生"白色污染"的制品。

（4）不乱扔垃圾及废弃物，将垃圾放到指定的垃圾箱内。提倡垃圾分类，清理"白色污染"，把废塑料袋打个结再扔进垃圾箱。

（5）支持和参与废纸、废玻璃、废塑料和废金属的回收利用，尽量减少生活垃圾。

（6）不焚烧一次性塑料制品，以免产生有害气体污染环境，看到焚烧垃圾的行为要勇敢的制止。

（7）在日常生活中看到随地乱扔的垃圾要随手捡起，放入垃圾箱。

（8）当好环保宣传员，带动家人和朋友使用环保产品。

（9）积极参加环保活动，做一名环保小卫士，为杜绝"白色污染"而努力。

---

* 南昌市乡村振兴局官方微信公众号，https://www.sohu.com/a/472787270_121106994.

4. 中国塑料污染治理体系与成效

经过几十年的努力，中国通过发展塑料循环经济，从全生命周期治理塑料污染，建成了世界上相对完善的塑料循环利用体系，治理成效逐渐显现。但同时，也面临着塑料消费刚性增长、塑料回收利用经济性下降等客观挑战。

4.1 中国塑料污染治理体系

为应对日渐严峻的塑料污染问题，中国政府出台了投资、财政、税收等一系列鼓励措施，规范和引导塑料废弃物回收利用，推动塑料行业绿色化、低碳化、循环化发展。在政府的推动下，企业和社会公众广泛参与，形成了覆盖广泛的回收利用体系。

4.1.1 制定出台塑料污染治理法律法规

法律法规是推动塑料污染治理的有效手段，对规范相关主体行为、构建塑料废弃物回收利用体系、推动塑料循环经济发展具有重要作用。

（1）将塑料污染治理纳入环境基本法中。

在《中华人民共和国环境保护法》中规定，国务院和沿海地方各级人民政府应当加强对海洋环境的保护，地方各级人民政府应当采取措施，组织对生活废弃物的分类处置、回收利用。在《中华人民共和国固体废物污染环境防治法》中规定，禁止任何单位或者个人向江河、湖泊、运河、渠道、水库及其最高水位线以下的滩地和岸坡以及法律法规规定的其他地点倾倒、堆放、贮存固体废物；电子商务、快递、外卖等行业应当优化物品包装，减少包装物的使用，并积极回收利用包装物；旅游、住宿等行业应当按照国家有关规定不主动提供一次性用品。

（2）出台重点领域塑料污染治理部门规章。

为加强重点领域塑料污染治理，中国政府有关部门出台了《邮件快件包装管理办法》，要求寄递企业优先采用可重复使用、易回收利用的包装物，优化邮件快件包装，减少包装物的使用，并积极回收利用包装物；出台了《农用薄膜管理办法》，对农用薄膜的生产、使用、回收和再生循环利用进行规范，强化生产者、使用者的回收责任，防止造成塑料污染；出台了《农药包装废弃物回收处理管理办法》，规定农药生产者、经营者应当按照"谁生产、经营，谁回收"的原则，履行相应的农药包装废弃物回收义务；出台了《废塑料加工利用污染防治管理规定》，对废塑料材料化再生利用进行规范，防止在废塑料再生循环利用过程中产生二次环境污染。

4.1.2 不断完善塑料污染治理标准体系

（1）完善塑料制品设计生产环节的标准。

中国政府有关部门制定了《生态设计产品评价通则》《电子电气生态设计产品评价通则》《生态设计产品标识》等一系列生态设计国家标准，规范含塑产品设计行为，减少塑料使用，提高废弃后的可回收性。"绿色再生塑料供应链联合工作组"牵头制定了《塑料制品易回收易再生设计评价通则》，从塑料制品的易回收、易再生性入手，通过产品设计与回收表现之间的关联关系来评价并指导塑料制品设计，以达到提高塑料回收率的目的。

（2）完善废塑料再生循环利用环节标准。

中国政府有关部门制定了《废塑料再生利用技术规范》，明确了废塑料再生利用过程的主要工艺环节，提出了生产加工过程的技术要求、助剂要求、装备要求，量化了废塑料再生利用关键环节的指标要求；制定发布了《塑料 再生塑料 第1部分：通则》等，对再生塑料通用质量要求和聚乙烯材料、聚丙烯材料等再生塑料质量标准等进行规范；制定发布了《废塑料回收与再生利用污染控制技术规范（试行）》，对废塑料的回收、贮存、运输、预处理、再生利用等过程中的环境保护相关事项提出了具体要求。

4.1.3 出台塑料污染治理激励政策措施

（1）制定塑料再生利用财政金融政策。

为支持和引导社会资金投入到塑料废弃物回收利用和塑料废弃物收集处置当中，中国政府对塑料废弃物再生循环利用项目与生活垃圾收集和焚烧项目建设给予适当的财政补贴，对利用塑料垃圾等生活垃圾焚烧发电产生的电力给予价格补贴，从而促进相关产业发展。同时，出台了《绿色债券发行指引》和《绿色信贷指引》，对包括塑料废弃物循环利用在内的环保项目给予优先发行债券等金融支持。

(2) 制定塑料再生利用税收优惠政策。

为促进再生塑料循环利用，中国政府出台了《资源综合利用产品和劳务增值税优惠目录》，对再生塑料产品给予70%的增值税即征即退优惠，对废农膜再生利用后的再生塑料制品及颗粒给予100%的增值税即征即退优惠；出台了《资源综合利用企业所得税优惠目录》，对企业生产目录内符合国家或行业相关标准的产品取得的收入，在计算应纳税所得额时，减按90%计入当年收入总额，以降低相关企业的所得税缴纳额度。

4.1.4 形成政府、企业和社会公众共同推进机制

(1) 政府负责制定规划和建设基础设施。

政府是塑料污染系统治理的推动者。中国政府有关部门先后制定了《关于建立完整的先进的废旧商品回收体系的意见》《关于加快推进再生资源产业发展的指导意见》等文件，推动包括废塑料在内的再生资源回收利用体系建设。在这一过程中，地方各级政府也积极推动建设完善的再生资源回收利用体系，实现了饮料瓶等塑料包装物的高效回收利用。此外，各地政府部门建设了完善的生活垃圾收集、转运和处置设施，建成了"村收集、镇转运、县处置"的城乡一体化收集处理体系，及时开展生活垃圾清运与处置，有效防止塑料废弃物泄漏到环境中。

(2) 企业全面参与塑料污染的全生命期治理。

企业是塑料污染治理的主要参与者和贡献者。在一次性塑料制品生产和使用环节，相关塑料制品企业按照国家有关规定，在淋洗类化妆品中停止使用塑料微珠，研发和创新包装物设计，减少一次性塑料制品的生产和使用；快递公司、餐饮饭店等广大流通和服务企业，积极遵守停止使用一次性塑料购物袋等禁限管理规定，积极开发绿色替代产品和包装物。在回收利用环节，企业积极建设塑料废弃物回收利用设施，加大再生塑料生产技术研发力度，对塑料废弃物进行合理的能源化利用，极大降低了其环境泄漏风险。

(3) 社会公众积极参与绿色消费和垃圾分类。

社会公众是塑料污染治理的践行者。在塑料污染治理过程中，越来越多的消费者主动选用环保产品、旅行时自带洗漱用品、拒绝过度包装商品，逐渐形成了

绿色消费新风尚。在日常生活中，消费者积极参与垃圾分类，不乱丢塑料垃圾，将可回收的饮料瓶等可回收塑料废弃物交给专门回收人员，助力塑料废弃物的再生循环利用。

4.2 中国塑料污染治理取得的成效

4.2.1 建成覆盖广泛、规模庞大的塑料废弃物回收体系

中国基于本国国情，不断建设完善塑料回收基础设施，建立了由回收网点、分拣中心、加工利用工厂组成的完善的废塑料回收利用体系。特别是近年来，中国大力推行垃圾分类，推动垃圾分类网络与再生资源回收利用网络"两网融合"，利用互联网、物联网等技术创新回收模式，有效促进了塑料废弃物的回收。另外，由于目前的塑料废弃物大多具有一定的经济价值，社会上存在大量自发从事塑料废弃物回收利用的个人和企业：一些个人以收购塑料废弃物等废弃物品为生，形成个体自发回收模式；一些专业回收企业面向消费者和个体回收人员开展回收业务，形成企业自主回收模式；一些回收企业依托环卫工人，利用环卫系统已有的分拣设施获得可再生资源，形成合作回收模式。

目前，中国废塑料回收与再生利用产能世界第一。根据最新调查数据，在中国从事废塑料回收和再生利用的企业数量超过了15000家，从事废塑料回收利用工作的从业人员规模约为90万人。[1]

---

[1] 再生塑料行业从业人员调研报告[R]. 中国物资再生协会再生塑料分会, 2022.

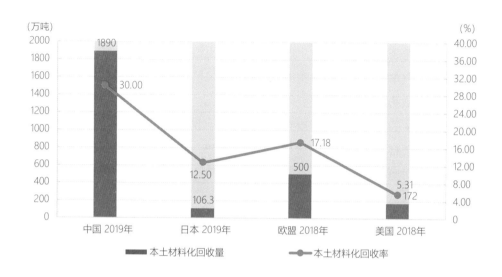

图4-1 中国与其他国家或地区本土材料化回收量和回收率对比

资料来源：《中国再生塑料行业发展报告》，中国物资再生协会再生塑料分会；《日本塑料产品、废弃物与资源回收2019》，日本塑料循环利用协会（PWMI）；《欧洲塑料生产、需求与废弃物数据分析报告（2020）》，PlasticsEurope；美国环境部网站（epa.gov）/美国统计局U.S. Census Bureau。

近年来，中国的废塑料回收量年均增长率稳定在2.5%左右。2021年，中国废塑料产生量约为6200万吨，其中材料化回收约为1900万吨，材料化回收率达到31%，是全球废塑料平均材料化回收率的近1.74倍[1]。而且，中国完全是在本国实现的材料化回收利用，如图4-1所示，2018年美国、欧盟，2019年日本本土材料化回收率分别只有5.31%、17.18%和12.50%，本土材料化再生利用总量是778万吨，中国2019年的材料化再生利用量是其2.43倍。据测算，2017年中国全面停止进口他国废塑料之前，中国废塑料回收利用产能约占全球的70%，2019年中国完全停止进口海外废塑料后，产能占比下降到了约58%，有大量中资废塑料回收利用企业转移到东南亚等地区，继续回收和加工利用来自欧、美、日的废塑料。[2]

---

[1] 中国再生塑料行业发展报告2020-2021[R].中国物资再生协会再生塑料分会，2021.

[2] 再生塑料行业从业人员调研报告[R].中国物资再生协会再生塑料分会，2022.

4.2.2 构建起完善的废塑料再生利用体系

中国依托庞大的塑料工业，建立起完善的废塑料再生利用体系。目前再生塑料在纺织、汽车、包装、消费类电子、农业、建筑建材等方面得到了广泛应用，几乎覆盖了从花盆、垃圾桶等低价值产品到家电、汽车等高价值产品大多数塑料工业领域。中国对不同类别的废弃塑料制品从回收利用价值上进行分类，对不同价值的废弃塑料制品采取不同的利用方式。例如，鼓励可重复使用产品和包装的循环利用，对塑料零部件和饮料瓶等开展物理再生，对膜袋类低值塑料废弃物进行能源化利用。

根据中国物资再生协会再生塑料分会统计分析，再生通用塑料在薄膜、注塑领域应用最广，占比分别达到36%和28%；在中空管材领域的应用占比达到12%。同时，再生工程塑料在纺织、汽车、包装、消费类电子等方面的应用不断增多，例如再生聚碳酸酯（Polycarbonate，PC）在家电、汽车、板材、合金行业应用较为常见；再生聚酰胺（Polyamide，PA）在汽车制造、电子电器、医疗器械及纺织行业等用量较大（见图4-2）。

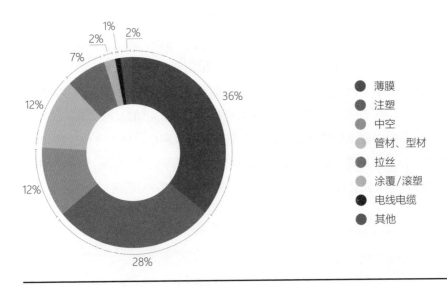

图4-2 2020年再生通用塑料下游应用情况
资料来源：根据《中国再生塑料行业发展报告2020-2021》整理绘制。

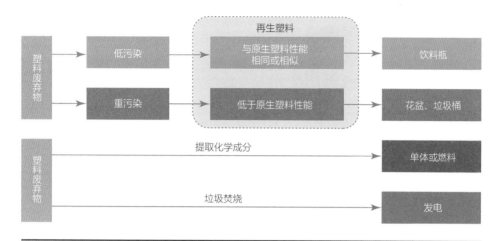

图4-3 废塑料梯次回收利用

通过多年来不断建设完善，中国目前已经构建起完善的塑料废弃物梯次利用体系（见图4-3）：首先是指采用物理再生的方法把塑料废弃物加工成与原生塑料性能相同或相近的再生塑料用于产品生产，比如废弃饮料瓶的"瓶到瓶"利用方式。这种利用方式可以最大化塑料废弃物的利用价值，实现"原级再生"；其次是对一些受污染较重、清洗难度大的废弃塑料制品，采用一定的加工方法加工成性能比原生塑料稍低的再生塑料，用于生产一些相对低端产品，比如花盆、垃圾桶等；再次是指对回收的塑料废弃物中的化学成分进行提取，使之成为单体或燃料，比如废塑料裂解制油等；最后是指通过垃圾焚烧直接利用废旧塑料中的能量用于焚烧发电。

2016年以来，中国城市生活垃圾焚烧设施逐年增多，日处理能力也日渐提升。如图4-4所示，城市生活垃圾焚烧设施从2016年的249座增加至2020年的463座，增长近1倍；焚烧处理能力从25.5850万吨/日提升至56.7804万吨/日，增长1.2倍。处理能力的提升显著提高了城市生活垃圾的焚烧处理量，5年间，焚烧处理量从7378万吨/年增长至14608万吨/年，年复合增长率约为30%。2020年，城市生活垃圾焚烧处理量占比已达到62.29%（见图4-4）。根据中国物资再生协会测算，其中塑料废弃物年能源化利用量达到2740万吨，能源化利用率为45.7%。

随着中国塑料回收利用产业的不断完善，废塑料回收利用产值也不断增加。

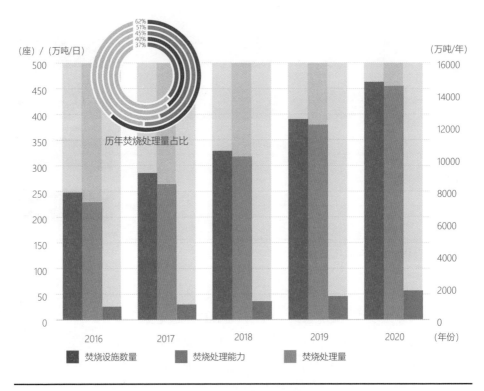

图4-4 2016~2020年中国城市（不包括县城）生活垃圾焚烧项目统计
注：历年焚烧处理量占比按实际处理量比例计算。
资料来源：住房和城乡建设部.城市建设统计年鉴（2016-2020）.

2019年，中国废塑料回收利用产值达到1000亿元；受新冠肺炎疫情及原料价格下跌等综合影响，2020年中国废塑料回收利用产值为790亿元，同比下降21%；2021年中国废塑料回收利用产值快速提升，达到1050亿元，同比增长33%。[1]

4.2.3 再生塑料利用有效减少了对化石原料的消耗

塑料废弃物具有资源和污染物双重属性，如果能有效回收利用就可以成为再生资源，如果处理不好才会成为污染物。因此，对塑料废弃物进行回收利用，实现"变废为宝"，不仅可以减少塑料污染，还可以为经济社会发展提供重要的资源，从而减少人类对石油等不可再生自然资源的过度消耗。

---

[1] 中国再生塑料行业发展报告2020-2021[R].中国物资再生协会再生塑料分会，2021.

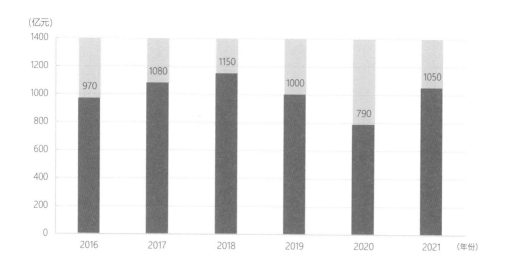

图 4-5 2016~2021 年中国废塑料回收价值
资料来源：中国物资再生协会再生塑料分会，商务部数据。

如图4-5和图4-6所示，2016~2021年的6年间，中国累计实现了1.08亿吨各类废塑料的材料化回收利用，回收价值超过6000亿元，如果按回收1吨废塑料相当于节约3吨石油计算，6年间中国累计节约石油开采和消耗3.3亿吨。[1] 通过促进废塑料循环利用，再生塑料已经成为中国资源的一个重要来源。与此同时，中国不仅对本国产生的塑料废弃物进行了有效的回收和再生利用，还处理利用了其他国家和地区大量塑料废弃物，为全球塑料废弃物的利用和处置做出了巨大贡献。据统计，自1992年以来，中国累计进口并材料化回收利用1.06亿吨废塑料，占全球同期废塑料材料化回收总量的45.1%[2]，成为全世界废塑料循环利用的中坚力量，相当于累计为世界节约了3.18亿吨原油的开采和消耗。

目前，中国正在大规模地开展生活垃圾分类，针对塑料污染治理出台了《"十四五"塑料污染治理行动方案》，以不断促进废塑料的回收和再生利用，覆盖各品类的废塑料循环利用体系将更加完善。如图4-6所示，预计2021~2025年中国废塑料材料化回收利用量将继续呈现稳步增长趋势，年度废塑料材料化回收

---

[1] 根据商务部数据统计整理。
[2] 根据世界主要国家公开数据及中国海关统计数据测算所得。

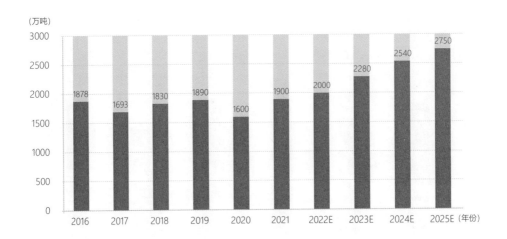

图 4-6 2016~2025 年中国废塑料回收量
注：2022E~2025E 为预测值。
资料来源：中国物资再生协会再生塑料分会。

利用量将保持在1900万~2750万吨之间，到2025年中国废塑料材料化回收利用量将达到2750万吨左右，材料化回收率将稳定在30%以上，如果加上能源化回收利用，整个塑料废弃物资源化利用率将稳定在75%以上。

可见，塑料回收利用不仅可以帮助解决塑料污染问题，还可以节约资源，减少对化石资源的使用，保障国家资源安全。

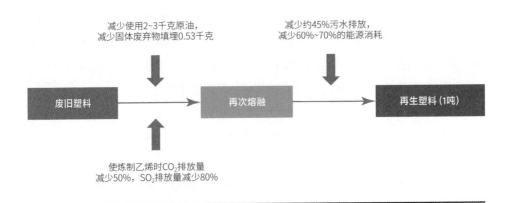

图 4-7 再生塑料生产过程中的减排情况
资料来源：邹琦志.废纸造纸企业减排措施研究[J].资源节约与环保,2014(6):33-34.

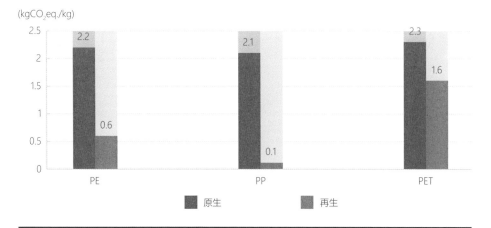

图 4-8 不同塑料生产过程的温室气体排放强度
资料来源：陈伟强，简小枚，王永刚，石磊.有关塑料的十大事实与再生塑料产业绿色发展建议[R].中国科学院城市环境研究所，2020.

4.2.4 为减少污染物和二氧化碳排放作出重要贡献

对废塑料进行回收利用，不仅可以避免由塑料废弃后本身带来的生态环境污染问题，与使用石油生产塑料相比，还有很好的污染物减排效果，可以使人类的发展建立在更加绿色环保的基础上。据测算，每回收利用1千克废塑料，相当于减少使用2~3千克原油，可减少固体废弃物产生0.53千克，可使炼制乙烯时二氧化碳（CO_2）排放量减少50%，二氧化硫（SO_2）排放量减少80%，可减少约45%的污水排放和60%~70%的能源消耗（见图4-7）[1]。2011~2020年10年间，中国累计实现1.7亿吨各类废塑料的材料化回收利用，相当于累计减少了5.1亿吨原油消耗，减少了0.9亿吨固体废物排放。

如图4-8所示，通过废塑料循环利用替代使用石油生产塑料，可以有效减少温室气体排放。据综合测算，废塑料回收利用温室气体减排效率约为0.36$kgCO_2eq./kg$废塑料，也就是说每利用1千克废塑料可以减少约0.36千克二氧化碳排放当量。[2] 据此，2011~2020年10年间，中国累计实现1.7亿吨各类废塑料的材料化回收利用，相当于减少了6120万吨二氧化碳排放，成为塑料领域实现"碳达峰、碳中和"的重要支撑。

---

[1] 戴铁军.包装废弃物的回收利用与管理[M].北京:科学技术出版社,2016.
[2] 邹琦志.废纸造纸企业减排措施研究[J].资源节约与环保,2014(6):33-34.

5. 中国塑料污染治理实践的启示和借鉴

经过几十年的不断努力，中国大力开展塑料污染全生命周期治理、绿色产业体系培育、绿色技术和商业模式创新、绿色消费倡导、多元共治等工作，逐步探索出了一条通过发展塑料循环经济解决塑料污染的有效路径。在当前中国大力推进生态文明建设、人民对美好生活的需要日益提高、全球2030年可持续发展议程加速推进的背景下，中国塑料污染治理的实践和经验对进一步完善和加强塑料污染治理、推动全球塑料污染治理合作，具有重要的启示和参考价值。

5.1 应对塑料污染需要建立完善的全生命周期管理体系

塑料污染治理是一项复杂的系统工程，涉及产品设计、生产、流通、消费、回收、处理、再生利用等诸多环节。中国塑料污染治理基于全生命周期管理理念，旨在构建一个从产品设计生产、消费使用到末端回收处置的完整治理体系。

（1）需要认识到一次性塑料制品的减量和替代只是全生命周期管理的一个环节。

当前，世界主要国家和地区针对塑料袋、塑料吸管等一次性塑料制品普遍出台了法律法规，加强消费领域一次性塑料制品的治理，这些措施对应对塑料污染是有益的，也是必要的。但必须清楚地认识到，一次性塑料制品在整个塑料工业体系中的占比较低，一次性塑料制品减量替代仅是塑料污染全生命周期治理的一部分，对塑料污染治理的贡献有限。

（2）完善的塑料废弃物末端收集处置基础设施是塑料污染防治的关键。

塑料污染治理需要建立起从塑料制品设计生产到末端处置利用的完善的管理体系，其中塑料废弃物的有效收集和处置设施的建设，可以直接防止塑料废弃物的环境泄漏。因此，各个国家和地区在塑料污染治理中应当优先对塑料废弃物收集和处置设施建设予以关注和投入。只有这样，才能建立起全球塑料污染防治的坚固堤坝。

5.2 发展塑料循环经济需要建立完善基于本国的回收利用体系

废塑料种类繁杂，利用和处置方式多样，需要针对各类废塑料建立完善的回收利用体系。在废塑料回收利用体系建设过程中，加强产品设计生产与回收利用、分类回收与分级利用的有效衔接，才能有效提高废塑料回收利用效率。

(1) 塑料废弃物回收体系需要不断完善和提升。

随着塑料制品消费的持续增加、塑料废弃物的产生量增加，以及城市化的快速发展，要逐步改变传统的以个体回收为主的回收模式，推动公司化、规范化回收，并积极采用互联网、智能回收机、智能分拣等新技术新装备，探索专业回收、企业整体回收等新型回收模式，提高回收效率、降低回收成本、加强废塑料流向管理，提高回收体系的现代化水平。

(2) 再生塑料合理利用是提高回收率的关键。

从全社会的角度看，一定时期内可进行材料化回收利用的废塑料总量及经济技术条件制约是一定的，因此，过度强调废塑料"原级利用"会显著增加废塑料的回收利用成本，并不能提高全社会废塑料再生利用的规模和环境效益。因此，应在合理的技术经济条件下，建设不同等级的废塑料再生利用体系，既可以实现某些品种的原级利用，也能够实现废塑料在木塑、快递袋等产品的梯级利用，为相关再生材料寻找最优的终端应用途径，有效提高全社会塑料废弃物材料化回收利用率。

(3) 塑料废弃物的分类回收应根据最终的处置利用方式开展。

对于电器电子产品、汽车等产品中含的塑料，应加强这些废弃产品的规范处置，确保将塑料分离出来并进行再生利用。对于饮料瓶、塑料餐盒等独立的塑料制品，应将其作为可回收物，建设完善的回收体系，实现这些废弃塑料制品的分类回收。对于受污染严重、难以分拣的混杂塑料，能源化利用仍是较好的方式，因此可以将其与其他生活垃圾一并收集处置，进行能源化利用或填埋，不需要进行分类回收。

5.3 发展塑料循环经济需要综合考虑环境效益与经济效益

发展塑料循环经济时只有兼顾环境效益和经济效益，才能实现可持续的发展。

（1）必须建立可持续的塑料循环经济发展模式。

塑料循环经济发展过程中，不能仅追求环境效益而忽视经济效益，没有经济效益无法激发利益相关者的积极性和自主性，而且这样的发展模式也是不可持续的。但是，也不能只重视经济效益而置环境效益于不顾，因为很多废塑料的回收利用是不具备经济价值的。这就需要通过征收基金、给予补贴等方式建立价值补偿机制，使其回收利用行为具有经济性。

（2）技术创新突破是实现废塑料回收利用的关键。

废塑料能否实现再生利用在很大程度上取决于回收利用技术的突破。再生塑料高值利用技术装备、绿色改性剂开发应用、废塑料再生过程污染防治、低值废塑料能源利用与污染防治等技术能显著提高废塑料再生利用水平，实现经济效益与环境效益的平衡，促进塑料循环经济发展。

（3）商业模式创新有助于实现经济与环境效益平衡。

回收模式的创新可以有效促进废塑料的回收，降低回收成本。比如，通过实施押金制激励终端消费者把废弃的农药瓶、饮料瓶等交到收集点，可以减少回收过程中的收集成本。通过采取"互联网+回收"的方式回收居民家中的废旧家电，实现在线预约，可以优化回收路线、提高回收效率，也有助于降低回收成本。

5.4 发展塑料循环经济需要科学比较分析各种塑料替代产品和方案

目前,出现了很多塑料污染治理的新材料、新技术和替代方案。比如,生物可降解材料、化学回收技术、纸质包装、可循环包装等。但每种新方案和新技术都需要开展全生命周期分析,确保不会为了解决塑料污染却带来新的污染,特别是会增加全社会污染物总量的方案更是不可取。

(1) 化学回收仍需进一步探索可持续的商业模式和产品方案。

目前,化学回收被认为是可以实现废塑料高品质回收利用的一种有效方案。但是与物理回收相比,当前条件下化学回收的成本要更高。因此,对于能够物理回收的品种,物理回收仍是优先选择。在无法实现物理回收的塑料袋、塑料膜等低值废塑料回收领域,目前化学回收技术仍需要在源头对塑料废弃物按材质进行分类,并探索合理可行的商业运行模式。另外,单纯以燃料油为产品的塑料废弃物回收利用技术路线,与塑料废弃物直接焚烧发电相比,其综合经济环境效益仍需开展科学的评价分析。

(2) 各种替代产品的使用需要开展科学的对比分析。

目前,为了减少消费领域一次性塑料的使用,出现了纸吸管、纸包装、竹木餐具等替代产品。然而各种塑料的替代产品在技术可行性、经济性、可推广性、适用性、碳排放量、环境影响等方面都还各有欠缺,尚不存在一种完美的塑料替代品。与塑料制品相比,使用替代产品和方案需要从经济性和废弃后处置环节的全生命周期环境影响方面进行科学的对比分析。

5.5 发展塑料循环经济需要加强政企合作和引导全社会广泛参与

政府是塑料循环经济发展政策的制定者和监督者，企业是塑料制品的生产者和使用者，公民是塑料制品的消费者，因此发展塑料循环经济、开展塑料污染治理需要汇聚政府、企业、公民的多方力量，建设多元治理体系。

（1）政府要做好塑料循环经济的规范者与推动者。

政府需要制定和颁布有关法律法规，制定产品生态设计标准，加强生产者责任延伸制度等长效机制建设，出台有利于激励塑料循环经济发展的财政、税收、金融等支持政策体系，并采取切实措施鼓励与支持废塑料回收体系和加工利用项目建设，鼓励生产企业更多地使用再生塑料作为原料，为塑料循环经济发展提供必要条件。此外，政府还要建设生活垃圾收集处理设施，并确保其有效运行。

（2）企业要做好塑料循环经济的引领者与执行者。

企业应广泛采取行动，实现绿色转型发展。塑料制品生产企业应主动减少塑料微珠的使用，开展塑料制品"易回收、易再生"设计；塑料制品使用企业要制订一次性塑料制品减量替代方案，用环保产品替代一次性塑料制品，并积极扩大循环中转袋、可降解塑料制品、可循环塑料制品的使用。塑料回收利用企业要加快建设完善的废塑料循环利用体系，创新回收利用技术和模式。

（3）广大社会公众要做好塑料循环经济的参与者。

消费者在享受塑料带来的生活便利的同时，应该认识到塑料泄漏对环境造成的污染，在生活中主动减少一次性塑料制品的使用，积极参与垃圾分类，形成绿色生活、绿色消费的新风尚。此外，广大社会组织和新闻媒体要成为塑料循环经济的宣传员和相关知识的普及者，提高社会各界对塑料问题的科学认识，营造全社会共同参与的良好氛围。

5.6 塑料污染治理需要不断加强国际合作

我们只有一个地球，人类命运共同体需要聚焦可持续生产生活，致力于实现与自然和谐共生，共建清洁美丽世界。塑料制品也会随着国际贸易在全球范围内流动，面对塑料污染任何国家都不可能独善其身，需要加强国际间的交流与合作，帮助欠发达国家和地区尽快建立塑料循环经济体系。

(1) **各国都应尽快采取措施建立本国的塑料回收利用处置体系。**

废塑料的回收利用价值本就不高，建立完善基于国内循环的回收利用体系，有助于降低成本、提高回收的经济效益，从而提升废塑料的回收利用率。因此，各个国家，特别是有条件的发达国家和地区，应逐步改变传统的将"塑料废弃物"简单收集后出口到其他国家和地区的做法，转而在国内或区域内实现塑料废弃物的就地再生利用和科学处置，从而有效避免由于塑料废弃物跨境转移带来的环境污染风险。

(2) **需要高度关注塑料回收利用基础设施薄弱的欠发达国家和地区。**

在全球人类命运共同体的背景下，一些欠发达国家和地区塑料废弃物收集处理基础设施非常不健全，塑料回收利用率相对较低，环境泄漏风险大，成为全球塑料污染治理的短板。因此，应禁止"塑料废弃物"向这些不具备更好回收利用条件的国家和地区出口。同时，发达国家和国际组织应当积极关注这些国家和地区，在资金、技术、管理、人才等方面给予适当帮助，共同推进全球塑料污染治理进程。

(3) **应加快构建全球塑料废弃物流向监测评估体系。**

塑料污染的全球治理，需要建立在科学的数据分析基础上。因此，需要加快构建全球范围内的塑料废弃物跨区域流向监测和评估体系，借助物质流分析方法，定量和动态追踪塑料废弃物在国际间的流动，对各个国家和地区的塑料废弃物处置利用进行有效评估，预测未来发展趋势，及时评估风险、发现问题，寻找解决方案，推动全球范围内塑料污染的科学治理、协同治理。

6.
加强全球塑料污染治理的倡议

当前,各国共同面临着日益严重的塑料污染问题,特别是海洋塑料污染问题日益成为世界各国普遍关注的焦点,给人类的可持续发展带来极大挑战。

根据联合国环境规划署2021年发布的报告显示，1950~2017年期间全球累计生产的约92亿吨塑料中，约有70亿吨成为塑料废弃物，塑料的回收率不到10%。全世界塑料废物年产生量约为3亿吨左右[1]，大量的塑料废物进入土壤和海洋，最终形成"白色污染"，给生态环境保护和生物多样性带来严重威胁。

塑料污染是人类面临的共同挑战，任何国家都不可能独善其身。因此需要树立人类命运共同体意识，全世界联合起来共同采取积极行动加以应对，形成各个国家和地区广泛参与的"塑料污染治理共同体"。

为此，我们提出如下倡议：

（1）我们认为，塑料污染的本质是塑料废弃后的环境泄漏，塑料本身不是污染物，塑料污染治理的重点是塑料废弃后的回收和处置。但同时，我们也意识到开展塑料污染全生命周期治理，是减少塑料末端处置压力的重要途径。

（2）我们观察到，全球塑料污染问题的形成具有历史累积性和跨区域转移特征，因此各国不仅需要立即采取行动对当前产生的塑料废弃物进行有效治理，还需要对历史形成的塑料废弃物采取有效措施加以管控。

（3）我们认为，全球塑料污染治理迫切需要各个国家和地区立即采取行动，建设完善基于本国的塑料废弃物收集处置和再生循环利用设施，防止塑料废弃物泄漏到自然环境中。

（4）我们认为，发展塑料循环经济应当优先鼓励废塑料的材料化回收利用，其次是能源化回收利用，最后是塑料废弃物的规范填埋处置，禁止塑料废弃物的随意丢弃。

（5）我们认为，塑料污染和治理需要开展广泛的国际合作，鼓励各个国家和地区，包括私营部门在内的所有利益相关方加强区域、国家和地方间的双边与多边合作。

---

[1] United Nations (UN) Environment. *Beat Plastic Pollution*[EB/OL]. https://www.unenvironment.org/interactive/beat-plastic-pollution/.

（6）我们认为，各国应当对塑料废弃物的跨境转移进行合理管控，出口国应当确保塑料废弃物的接收国具备完善的处置利用基础设施和条件，能够避免产生二次污染，并在必要时对接收国提供支持或帮助。

（7）我们倡导，各个国家和地区都应根据本国国情制定出台有关塑料污染治理的法律法规和行动计划，提出塑料污染治理的阶段目标，并为之采取切实有效的行动。

（8）我们认为，有效应对全球塑料污染治理，需要发达国家和地区在基础设施建设及管理能力提升等方面，对欠发达国家和地区给予必要的资金、技术、人才等支持。

（9）我们认为，塑料污染，特别是海洋塑料污染需要建立科学的监测与评估体系，对塑料污染的形成和跨区域流动进行科学评判，指导全球塑料污染治理的科学开展。

（10）我们认为，全球塑料污染治理需要及时总结各国成功的经验及做法，形成可供各个国家和地区参考的案例与行动指南，提升各个国家和地区的塑料污染治理综合能力。

结束语

当前,塑料污染已经成为全球仅次于气候变化的焦点环境问题,引起了各个国家和地区的高度重视。中国经过几十年的不断探索和努力,探索出了一条通过发展塑料循环经济解决塑料污染治理的"中国方案",为全球塑料污染治理做出了重要贡献。但我们也应该看到,塑料污染治理是一个复杂的系统工程,仍存在许多不足和需要改进提升的地方,需要更多的国家和地区加入,塑料污染治理仍任重道远。

但我们也坚信,只要世界各国共同合作、积极参与,在不久的将来塑料污染问题一定会得到有效解决,人与自然和谐共生的可持续发展愿景必将实现!

参考文献

[1] 陈伟强,简小枚,王永刚,石磊.有关塑料的十大事实与再生塑料产业绿色发展建议[R].中国科学院城市环境研究所,2020.

[2] 戴铁军.包装废弃物的回收利用与管理[M].北京:北京科学技术出版社,2016.

[3] 工业和信息化部、发展和改革委员会、环境保护部.关于开展工业产品生态设计的指导意见[R].工信部节能与综合利用司,2013.

[4] 韩立钊,王同林,姚燕."白色污染"的污染现状及防治对策研究[J].中国人口·资源与环境,2010,20(S1):402-404.

[5] 侯彩霞.超临界水降解塑料的研究[D].天津大学,2003.

[6] 限塑一年全国超市塑料袋减少了2/3[EB/OL]. 华西都市报, http://news.ifeng.com.mainland/special/2010lianghui/zuixin/201003/0311-9417-1571533-1.shtml,2010/03/11.

[7] 马占峰,姜宛君.中国塑料加工工业(2020)[J].中国塑料,2021,35(5):119-125.

[8] 日本塑料产品、废弃物与资源回收2019[R].日本塑料循环利用协会(PWMI),2019.

[9] 微塑料渗入生态系统危及健康[J].橡塑技术与装备,2017,43(8):67-68.

[10] Wangliming.废旧家电塑料回收与利用技术的发展 [EB/OL].https://zhuanlan.zhihu.com/p/354754130,2021/3/5.

[11] 一年6000万吨,我国废旧塑料循环利用应从哪些方面入手?[EB/OL].https://www.sohu.com/a/479847539_1211 23882,2021/7/27.

[12] 再生塑料行业从业人员调研报告[R].中国物资再生协会再生塑料分会,2022.

[13] 郑强.塑料与"白色污染"刍议[EB/OL].https://mp.weixin.qq.com/s/qLPyAVI2xt49-QiUCOHdPA,2021/11/19.

[14] 中国快递包装废弃物产生特征与管理现状研究报告[R].绿色和平,摆脱塑缚,2019年11月.

[15] 中国再生塑料行业发展报告2019-2020[R].中国物资再生协会再生塑料分会,2020.

[16] 中国再生塑料行业发展报告2020-2021[R].中国物资再生协会再生塑料分会,2021.

[17] 邹琦志.废纸造纸企业减排措施研究[J].资源节约与环保,2014(06):33-34.

[18] Andrady A.L. Microplastics in the marine environment[J]. *Marine Pollution Bulletin*, 2011(62):1596-1605.

[19] Dees J. P., Ateia M., Sanchez D. L. Microplastics and their degradation products in surface waters: A missing piece of the global carbon cycle puzzle[J]. *ACS ES&T Water*, 2020,1:214-216.

[20] Dooley E. E. The Center for International Environmental Law[J]. *Environmental Health Perspectives*, 2004, 112(2):A91-A91.

[21] Hoornweg D., Bhada-Tata P., Kennedy C. Environment: Waste production must peak this century[J]. *Nature*,2013,502:615–617.

[22] Horton A. A., Walton A., Spurgeon D. J., et al.. Microplastics in freshwater and terrestrial environments: Evaluating the current understanding to identify the knowledge gaps and future research priorities[J]. *Science of the Total Environment*, 2017,586:127-141.

[23] Landrigan P. J., Stegeman J., Fleming L., Allemand D., Anderson D., Backer L., et al.. Human health and ocean pollution[J]. *Annals of Global Health*,2020,1:1-64.

[24] Lau W., Shiran Y., Bailey R. M., et al.. Evaluating scenarios toward zero plastic pollution[J]. *Science* (New York, N. Y.), 2020,6510(369):1455-1461.

[25] Macleod M., Arp H., Tekman M. B., et al.. The global threat from plastic pollution[J]. *Science* (New York, N. Y.), 2021,6550(373):61-65.

[26] Nizzetto L., Futter M., Langaas S. Are agricultural soils dumps for microplastics of urban origin? [J] *Environmental Science & Technology*, 2016, 50:10777-10779.

[27] *Plastic & Climate: The Hidden Costs of a Plastic Planet*[R]. Center for International Environmental Law, Environmental Integrity Project,2019.

[28] *Plastics-the Facts 2020: An Analysis of European Plastics Production, Demand and Waste Data*[R]. Plastics Europe Association of Plastics Manufactures, 2020.

[29] *From Pollution to Solution : A Global Assessment of Marine Litter and Plastic Pollution*[R]. United Nations Environment Programme, 2021.

[30] *The New Coal Plastics & Climate Change*[R]. America: Bennington College, Beyond Plastics, October2021.

[31] United Nations (UN) Environment. *Beat Plastic Pollution*[EB/OL]. https://www.unenvironment.org/interactive/beat-plastic-pollution/.

[32] Yu Y., Yang J., Wu W. M., et al.. Biodegradation and Mineralization of Polystyrene by Plastic-Eating Mealworms: Part2. Role of Gut Microorganisms[J]. *Environmental Science & Technology*, 2015,49(20):12087-12093.

附录
中国历年塑料污染治理相关法规政策文件汇总摘编

类别	制修订时间	政策文件	核心内容
法律法规	1989年12月制定 2014年4月修订	《中华人民共和国环境保护法》	国务院和沿海地方各级人民政府应当加强对海洋环境的保护。向海洋排放污染物、倾倒废弃物，进行海岸工程和海洋工程建设，应当符合法律法规规定和有关标准，防止和减少对海洋环境的污染损害。 国家鼓励和引导公民、法人和其他组织使用有利于保护环境的产品和再生产品，减少废弃物的产生。 地方各级人民政府应当采取措施，组织对生活废弃物的分类处置、回收利用。
	1995年10月制定 2020年4月修订	《中华人民共和国固体废物污染环境防治法》	禁止任何单位或者个人向江河、湖泊、运河、渠道、水库及其最高水位线以下的滩地和岸坡以及法律法规规定的其他地点倾倒、堆放、贮存固体废物。 县级以上地方人民政府应当加快建立分类投放、分类收集、分类运输、分类处理的生活垃圾管理系统，实现生活垃圾分类制度有效覆盖。 任何单位和个人都应当依法在指定的地点分类投放生活垃圾。禁止随意倾倒、抛撒、堆放或者焚烧生活垃圾。 产生秸秆、废弃农用薄膜、农药包装废弃物等农业固体废物的单位和其他生产经营者，应当采取回收利用和其他防止污染环境的措施。 电子商务、快递、外卖等行业应当优先采用可重复使用、易回收利用的包装物，优化物品包装，减少包装物的使用，并积极回收利用包装物。 国家依法禁止、限制生产、销售和使用不可降解塑料袋等一次性塑料制品。 国家鼓励和引导减少使用、积极回收塑料袋等一次性塑料制品，推广应用可循环、易回收、可降解的替代产品。 旅游、住宿等行业应当按照国家有关规定推行不主动提供一次性用品。

续表

类别	制修订时间	政策文件	核心内容
法律法规	2008年8月	《中华人民共和国循环经济促进法》	生产列入强制回收名录的产品或者包装物的企业，必须对废弃的产品或者包装物负责回收；对其中可以利用的，由各该生产企业负责利用；对因不具备技术经济条件而不适合利用的，由各该生产企业负责无害化处置。 对列入强制回收名录的产品和包装物，消费者应当将废弃的产品或者包装物交给生产者或者其委托回收的销售者或者其他组织。 从事工艺、设备、产品及包装物设计，应当按照减少资源消耗和废物产生的要求，优先选择采用易回收、易拆解、易降解、无毒无害或者低毒低害的材料和设计方案，并应当符合有关国家标准的强制性要求。 餐饮、娱乐、宾馆等服务性企业，应当采用节能、节水、节材和有利于保护环境的产品，减少使用或者不使用浪费资源、污染环境的产品。 国家在保障产品安全和卫生的前提下，限制一次性消费品的生产和销售。具体名录由国务院循环经济发展综合管理部门会同国务院财政、环境保护等有关主管部门制定。对列入前款名录中的一次性消费品的生产和销售，由国务院财政、税务和对外贸易等主管部门制定限制性的税收和出口等措施。 国家鼓励和推进废物回收体系建设。地方人民政府应当按照城乡规划，合理布局废物回收网点和交易市场，支持废物回收企业和其他组织开展废物的收集、储存、运输及信息交流
部门规章	2020年4月	《农用薄膜管理办法》	禁止生产、销售、使用国家明令禁止或者不符合强制性国家标准的农用薄膜。鼓励和支持生产、使用全生物降解农用薄膜。 农用薄膜使用者应当在使用期限到期前捡拾田间的非全生物降解农用薄膜废弃物，交至回收网点或回收工作者，不得随意弃置、掩埋或者焚烧。 农用薄膜生产者、销售者、回收网点、废旧农用薄膜回收再利用企业或其他组织等应当开展合作，采取多种方式，建立健全农用薄膜回收利用体系，推动废旧农用薄膜回收、处理和再利用

续表

类别	制修订时间	政策文件	核心内容
部门规章	2020年8月	《农药包装废弃物回收处理管理办法》	农药生产者（含向中国出口农药的企业）、经营者和使用者应当积极履行农药包装废弃物回收处理义务，及时回收农药包装废弃物并进行处理。 农药生产者、经营者应当按照"谁生产、经营，谁回收"的原则，履行相应的农药包装废弃物回收义务。 农药经营者应当在其经营场所设立农药包装废弃物回收装置，不得拒收其销售农药的包装废弃物。 农药使用者应当及时收集农药包装废弃物并交回农药经营者或农药包装废弃物回收站（点），不得随意丢弃。 国家鼓励和支持对农药包装废弃物进行资源化利用；资源化利用以外的，应当依法依规进行填埋、焚烧等无害化处置。 农药包装废弃物处理费用由相应的农药生产者和经营者承担；农药生产者、经营者不明确的，处理费用由所在地的县级人民政府财政列支
	1990年11月	《食品用塑料制品及原材料卫生管理办法》	凡加工塑料食具、容器、食品包装材料，不得使用回收塑料
	2005年12月	《产业结构调整指导目录》（2005年本）	鼓励复合材料和高分子材料，淘汰泡沫塑料及一次性发泡塑料餐具
	2012年8月	《废塑料加工利用污染防治管理规定》	禁止利用废塑料生产厚度小于0.025mm的超薄塑料购物袋和厚度小于0.015mm超薄塑料袋。 废塑料加工利用单位应当以环境无害化方式处理废塑料加工利用过程产生的残余垃圾、滤网；禁止交不符合环保要求的单位或个人处置。禁止露天焚烧废塑料及加工利用过程产生的残余垃圾、滤网。 废塑料加工利用集散地应当建立废塑料加工利用散户产生的残余垃圾和滤网集中回收处理机制。鼓励废塑料加工利用集散地对废塑料加工利用散户实行集中区化管理，集中处理废塑料加工利用产生的废水、废气和固体废物

续表

类别	制修订时间	政策文件	核心内容
规范性文件	1989年9月	《关于加强重点交通干线、流域及旅游景区塑料包装废物管理的若干意见》	禁止在铁路车站和旅客列车、长江及太湖等内河水域航运的客船和旅游船上使用不可降解的一次性发泡塑料餐具。 杜绝塑料包装废物及其他固体废物在河流、湖泊中及沿岸乱扔和堆积。长江、太湖、重点旅游景区（景点）和其他内河水域，已经在水中漂浮和岸边堆积的塑料包装废物，三个月内在辖区地方人民政府统一领导下负责组织清理干净。 禁止在铁路沿线、长江、太湖流域沿岸倾倒垃圾。 各类船舶要依据有关法律配备足够的垃圾储存容器，对垃圾分类收集，并排入垃圾接收设施。禁止船员和乘客向江（湖）中抛弃垃圾和货物残余物等。 各级旅游景区（景点）的主管部门负责对所辖景区（景点）塑料包装废物的管理进行监督和检查。各景区（景点）应配备足够的垃圾收集容器，方便游客投放垃圾。旅游景区（景点）的管理单位要设置专人清扫和收运垃圾，维护垃圾收贮设施
	2001年4月	《关于立即停止生产一次性发泡塑料餐具的紧急通知》	所有生产企业(包括国内投资、外商投资和港、澳、台商投资企业)要自觉遵守国家法律法规和贯彻执行国家产业政策，立即停止生产一次性发泡塑料餐具
	2002年1月	《关于加强淘汰一次性发泡塑料餐具执法监督工作的通知》	各地工商、质检、环保等执法部门要切实负责，从本通知印发之日起，依法加强对本地区淘汰一次性发泡塑料餐具工作的监督检查
	2007年12月	《国务院办公厅关于限制生产销售使用塑料购物袋的通知》	从2008年6月1日起，在全国范围内禁止生产、销售、使用厚度小于0.025毫米的塑料购物袋(以下简称超薄塑料购物袋)。 自2008年6月1日起，在所有超市、商场、集贸市场等商品零售场所实行塑料购物袋有偿使用制度，一律不得免费提供塑料购物袋
	2009年1月	《国务院办公厅关于治理商品过度包装工作的通知》	要在满足保护、保质、标识、装饰等基本功能的前提下，按照减量化、再利用、资源化的原则，从包装层数、包装用材、包装有效容积、包装成本比重、包装物的回收利用等方面，对商品包装进行规范，引导企业在包装设计和生产环节中减少资源消耗，降低废弃物产生，方便包装物回收再利用

续表

类别	制修订时间	政策文件	核心内容
规范性文件	2010年5月	《国家发展改革委、财政部关于开展城市矿产示范基地建设的通知》	通过5年的努力,在全国建成30个左右技术先进、环保达标、管理规范、利用规模化、辐射作用强的"城市矿产"示范基地。推动报废机电设备、电线电缆、家电、汽车、手机、铅酸电池、塑料、橡胶等重点"城市矿产"资源的循环利用、规模利用和高值利用
	2011年10月	《国务院办公厅关于建立完整的先进的废旧商品回收体系的意见》	充分发挥市场机制作用,提高废金属、废纸、废塑料、报废汽车及废旧机电设备、废轮胎、废弃电器电子产品、废玻璃、废铅酸电池、废弃节能灯等主要废旧商品的回收率。加强政策引导和支持力度,进一步明确生产者、销售者、消费者责任,通过垃圾分类回收等途径,切实做好重点废旧商品的有效回收
	2013年1月	《国务院关于印发循环经济发展战略及近期行动计划的通知》	落实有关优惠政策,做好废金属、废塑料、废玻璃、废纸等传统再生资源的回收,提高回收率。 鼓励自备购物袋,禁止使用超薄塑料购物袋
	2013年4月	《关于深化限制生产销售使用塑料购物袋实施工作的通知》	发展改革部门会同有关部门通过电视、网络、广播、报纸等媒体,采取多种形式大力宣传"限塑令"实施以来在节约能源资源、提升环保意识等方面取得的积极成效,倡导绿色、低碳、节约的消费理念。 教育部门要在中小学生节约教育和环境教育中宣传减少使用塑料购物袋等一次性制品,倡导中小学生在日常行为中坚持节约环保理念。 商务部门、价格部门、工商部门组织商场、超市、集贸市场开展"限塑令"宣传活动,号召消费者自觉抵制超薄塑料购物袋,推动经营者自觉落实有偿使用制度。商务部门组织相关协会、企业发起减少塑料购物袋使用倡议活动。 工业和信息化部门协调电信运营商在"限塑令"实施五周年期间向本地手机用户发送关于"限塑令"实施重要意义的温馨提示短信。 机关事务管理部门要在"限塑令"实施五周年期间,组织公共机构开展"限塑令"宣传活动,倡导机关工作人员率先垂范,减少使用塑料购物袋。 环保部门要结合"世界环境日"等活动大力宣传超薄塑料购物袋带来的环境问题,使消费者认清超薄塑料购物袋的危害,使不用、少用塑料购物袋成为一种自觉行为。 妇联组织通过城乡"妇女之家"等平台,采取丰富多彩、群众喜闻乐见的方式,宣传"限塑令"的重要意义以及白色污染对环境和人体健康的危害,倡导"拎起菜篮子"、"提起布袋子"

续表

类别	制修订时间	政策文件	核心内容
规范性文件	2015年12月	《废塑料综合利用行业规范条件》《废塑料综合利用行业规范条件公告管理暂行办法》	明确了行业新建、已建的三大重点类型企业在废塑料处理能力上的门槛
	2016年12月	《国务院办公厅关于印发生产者责任延伸制度推行方案的通知》	开展生态设计。生产企业要统筹考虑原辅材料选用、生产、包装、销售、使用、回收、处理等环节的资源环境影响，深入开展产品生态设计。具体包括轻量化、单一化、模块化、无（低）害化、易维护设计，以及延长寿命、绿色包装、节能降耗、循环利用等设计。 使用再生原料。在保障产品质量性能和使用安全的前提下，鼓励生产企业加大再生原料的使用比例，实行绿色供应链管理，加强对上游原料企业的引导，研发推广再生原料检测和利用技术。 规范回收利用。生产企业可通过自主回收、联合回收或委托回收等模式，规范回收废弃产品和包装，直接处置或由专业企业处置利用。产品回收处理责任也可以通过生产企业依法缴纳相关基金、对专业企业补贴的方式实现。 加强信息公开。强化生产企业的信息公开责任，将产品质量、安全、耐用性、能效、有毒有害物质含量等内容作为强制公开信息，面向公众公开；将涉及零部件产品结构、拆解、废弃物回收、原材料组成等内容作为定向公开信息，面向废弃物回收、资源化利用主体公开。 支持饮料纸基复合包装生产企业、灌装企业和循环利用企业按照市场化原则组成联盟，通过灌装企业销售渠道、现有再生资源回收体系、循环利用企业自建网络等途径，回收废弃的饮料纸基复合包装
	2016年12月	《关于加快推进再生资源产业发展的指导意见》	大力推进废塑料回收利用体系建设，支持不同品质废塑料的多元化、高值化利用。以当前资源量大、再生利用率高的品种为重点，鼓励开展废塑料重点品种再生利用示范，推广规模化的废塑料破碎—分选—改性—造粒先进高效生产线，培育一批龙头企业。积极推动低品质、易污染环境的废塑料资源化利用，鼓励对生活垃圾塑料进行无污染的能源化利用，逐步减少废塑料填埋
	2017年4月	《循环发展引领行动》	制定发布限制生产和销售的一次性消费品名录及管理办法，对纳入目录的产品实行分类管理，制定完善限制一次性消费品的相关政策。支持研发可重复使用的替代产品。研究制定一次性产品的生态设计标准，提高回收利用率

续表

类别	制修订时间	政策文件	核心内容
规范性文件	2017年7月	《国务院办公厅关于印发禁止洋垃圾入境推进固体废物进口管理制度改革实施方案的通知》	2017年底前，禁止进口生活来源废塑料、未经分拣的废纸以及纺织废料、钒渣等品种
	2018年12月	《国务院办公厅关于印发"无废城市"建设试点工作方案的通知》	以回收、处理等环节为重点，提升废旧农膜及农药包装废弃物再利用水平。建立政府引导、企业主体、农户参与的回收利用体系。 限制生产、销售和使用一次性不可降解塑料袋、塑料餐具，扩大可降解塑料产品应用范围。加快推进快递业绿色包装应用。 在宾馆、餐饮等服务性行业，推广使用可循环利用物品，限制使用一次性用品
	2020年1月	《国家发展改革委 生态环境部关于进一步加强塑料污染治理的意见》	禁止生产和销售厚度小于0.025毫米的超薄塑料购物袋、厚度小于0.01毫米的聚乙烯农用地膜。禁止以医疗废物为原料制造塑料制品。禁止生产和销售一次性发泡塑料餐具、一次性塑料棉签；禁止生产含塑料微珠的日化产品。 商场、超市、药店、书店等场所以及餐饮打包外卖服务和各类展会活动，禁止使用不可降解塑料袋，集贸市场规范和限制使用不可降解塑料袋。 餐饮行业禁止使用不可降解一次性塑料吸管；地级以上城市建成区、景区景点的餐饮堂食服务，禁止使用不可降解一次性塑料餐具。 星级宾馆、酒店等场所不再主动提供一次性塑料用品，可通过设置自助购买机、提供续充型洗洁剂等方式提供相关服务。 邮政快递网点禁止使用不可降解的塑料包装袋、塑料胶带、一次性塑料编织袋等。 在商场、超市、药店、书店等场所，推广使用环保布袋、纸袋等非塑制品和可降解购物袋。在餐饮外卖领域推广使用符合性能和食品安全要求的秸秆覆膜餐盒等生物基产品、可降解塑料袋等替代产品。在重点覆膜区域，结合农艺措施规模化推广可降解地膜。 积极推广可循环、可折叠包装产品和物流配送器具。 结合实施垃圾分类，加大塑料废弃物等可回收物分类收集和处理力度，禁止随意堆放、倾倒造成塑料垃圾污染。 开展江河湖泊、港湾塑料垃圾清理和清洁海滩行动。推进农田残留地膜、农药化肥塑料包装等清理整治工作，逐步降低农田残留地膜量

续表

类别	制修订时间	政策文件	核心内容
规范性文件	2020年7月	《关于扎实推进塑料污染治理工作的通知》	发布《相关塑料制品禁限管理细化标准（2020年版）》
	2021年9月	《"十四五"塑料污染治理行动方案》	积极推动塑料生产和使用源头减量，包括积极推行塑料制品绿色设计、持续推进一次性塑料制品使用减量、科学稳妥推广塑料替代产品等。 加快推进塑料废弃物规范回收利用和处置，包括加强塑料废弃物规范回收和清运、建立完善农村塑料废弃物收运处置体系、加大塑料废弃物再生利用、提升塑料垃圾无害化处置水平等。 大力开展重点区域塑料垃圾清理整治，有针对性地部署了江河湖海、旅游景区、农村地区的塑料垃圾清理整治任务

重点支持单位和企业

中国石油和化学工业联合会
中国物资再生协会
北京三快在线科技有限公司（美团）
金发科技股份有限公司
联想（北京）有限公司
玛氏食品（中国）有限公司
陶氏化学（中国）有限公司
埃克森美孚（中国）投资有限公司

致谢

本报告在编写过程中，荀彬、田瑾、高杨、龚勋、俞昕、侯聪、鲍晨、黄培坤、赵子康、王妍等相关专家学者，在资料收集、案例调研、报告编写等过程中给予了大力帮助，为本报告的形成发挥了重要作用，在此表示诚挚的感谢！

同时，本报告在编写过程中参考了大量的国内外文献、报告，感谢相关研究人员在塑料污染治理工作中开展的卓有成效的工作，为本研究报告的编写提供了重要支撑！

Report of the National Think Tank

Plastic Pollution Prevention and Control in China
Principles and Practice
(Extracted version)

1.
Urgency of Strengthening Plastic Pollution Control

The invention of plastic is one of the symbols of modern industrial civilization, bringing great convenience to human production and daily life. However, the plastic pollution issue is becoming increasingly prominent, causing environmental concerns across the globe.

1.1 Global Pollution Issues Arising from Plastics
are Increasingly Severe

Pursuant to the data from the Plastics Europe, global plastic production and consumption grew steadily at an average annual rate of 2% from 2015 to 2020. Plastic production is expected to double by 2035 and treble by 2050[1] (see Figure 1-1 for details).

According to a report released by the United Nations Environment Programme in 2021, approximately 9.2 billion tons of plastics were produced globally cumulatively between 1950 and 2017. By 2050, cumulative global plastic production is expected to grow to 34 billion tons, with an average annual growth rate of 7.9% (see Figure 1-2).[2]

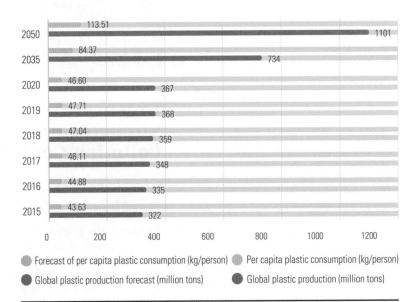

Figure 1-1 *Global Plastic Production and Consumption*
Source: Plastics Europe.

---

1. Plastics Europe. https://plasticseurope.org/knowledge-hub/.
2. From Pollution to Solution: A Global Assessment of Marine Litter and Plastic Pollution[R]. United Nations Environment Programme, 2021.

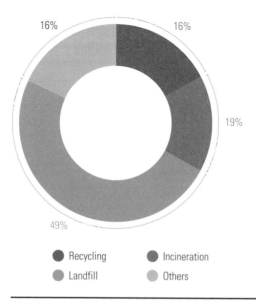

Figure 1-2 Composition of Global Plastic Disposal Methods
Source: Data from OECD.

In 2019, some 350 million tons of plastic waste were generated globally.[1] Meanwhile, the discharge of plastic waste continues to grow year by year. The global prevalence of serious plastic pollution problems urgently requires countries around the world to take extensive and effective measures to respond.

---

1. OECD, https://stats.oecd.org/Index.aspx#.

China is also Faced
with the Plastic Pollution Concern

Although as a developing country, China's per capita plastic consumption is much lower than that of developed countries, it is still facing the same plastic pollution issue.

1.2.1 Plastic consumption is increasing year by year in China

With China's economic development and rising living and consumption levels, the consumption of plastics continues to grow. In addition, changes in people's consumption patterns and the rapid development of emerging sectors in recent years, as well as the massive increase in demand for disposable medical and protective supplies due to the New Crown epidemic have also led to rapid growth in the use of disposable plastic products. In 2020, China's plastic consumption reached 98.77 million tons, up 12.2% year-on-year. [1]

1.2.2 The pressure imposed on China concerning plastic pollution control is increasingly high

As China's economy and people's consumption continue to grow, so does the consumption of plastics and the demand for plastic products. The inevitable result of the rising plastic consumption is the gradual increase of plastic waste. As shown in Figure 1-3, in recent years, China has produced more than 60 million tons of plastic waste annually. In addition to materialization and energy utilization, the remaining was basically sent to landfills together with other domestic waste, bringing certain risks to ecological and environmental safety.

---

1. *Report of Plastic Recycling Industry in China (2020-2021)* [R]. Recycled Plastics Association, China National Resources Recycling Association, 2021.

Chapter 1
Urgency of Strengthening Plastic Pollution Control

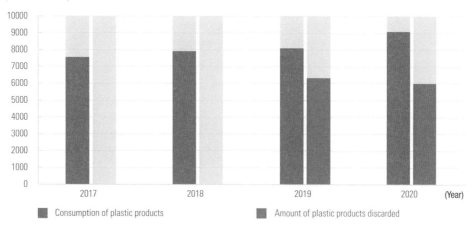

Figure 1-3 2017-2020 Consumption and Discarded Amount of Plastic Products in China
Note: Compiled from the annual series of articles "China Plastics Processing Industry (2020)" published in the journal *China Plastics* by Ma Zhanfeng et al.
Source: Based on *Report of Plastic Recycling Industry in China (2020-2021)*.

2.

The Nature and Manifestations of Plastic Pollution

Once plastic waste leaks into the natural environment, it will take hundreds or even thousands of years to completely degrade without human intervention, resulting into long-term adverse effects on the global soil environment, water ecology, climate change, biodiversity, etc.

 The Nature of Plastic Pollution

Currently, plastic is a prominently fundamental material in society, bringing great convenience to people's daily life. The plastic itself is not a pollutant. The nature of plastic pollution is the leakage into the natural environment caused by improper control of plastic waste. The current severe plastic pollution is the result of long-term accumulation in history.

2.1.1 The nature of plastic pollution is plastic waste leakage into the nature environment

In 2019, the global volume of plastics is equivalent to 26% of steel and cement combined, and its consumption is growing at an average annual rate of 2%.[1] Plastic has become an essential and important raw material. Similar to other industrial materials such as steel and non-ferrous metals, plastics are inherently highly recyclable and are theoretically fully recyclable to avert leakage.

However, due to its wide application in different areas with various forms, certain categories of plastic products could be easily discarded after using but difficult to collect, with the risk of leakage into the natural environment such as water and soil. Moreover, with high corrosion resistance quality, even a small plastic straw can stay in natural conditions for a long time and brings out serious pollution after years' accumulation. Compared with the environmental problems caused by other materials, the management of plastic pollution is more complicated.

---

1. Zheng Qiang. *Discussion on Plastics and "White pollution"* [EB/OL]. https://mp.weixin.qq.com/s/qLPyAVI2xt49-QiUCOHdPA,2021/11/19.

2.1.2 Global plastic pollution has been chronically accumulated for decades

The plastic industry commenced its rapid development in the 1960s and 1970s, leading to the spike in global plastic production and consumption. However, the concept of sustainable development has not yet become a global consensus, and no related policies, regulations, and control measures for plastic waste were established, resulting in wide plastic waste leakage in a long period.

With legislation and regulations in place by various nations, wider promotion of the eco-design of products, improvement of waste management system and recycling infrastructure, and the application of recycling and utilization technologies, we believe that human beings are able to impose effective control over plastic pollution.

2.2 The Manifestations of Plastic Pollution

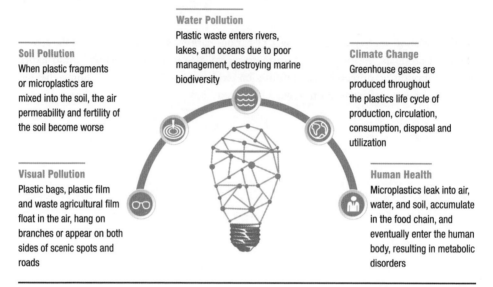

Soil Pollution
When plastic fragments or microplastics are mixed into the soil, the air permeability and fertility of the soil become worse

Water Pollution
Plastic waste enters rivers, lakes, and oceans due to poor management, destroying marine biodiversity

Climate Change
Greenhouse gases are produced throughout the plastics life cycle of production, circulation, consumption, disposal and utilization

Visual Pollution
Plastic bags, plastic film and waste agricultural film float in the air, hang on branches or appear on both sides of scenic spots and roads

Human Health
Microplastics leak into air, water, and soil, accumulate in the food chain, and eventually enter the human body, resulting in metabolic disorders

Figure 2-1 The Manifestations of Plastic Pollution

Once plastic waste leaks into the soil, water, and other natural environments, it is difficult to degrade, causing visual pollution, soil pollution, water pollution, and other environmental damage (see Figure 2-1), ending with permanent harm to the fragile ecological environment and biodiversity.

2.2.1 Visual pollution caused by plastic waste

The visual pollution of plastic waste refers to the damage caused by plastic waste scattered in the environment and natural landscape of the city.[1] Plastic bags, packaging films, waste agricultural films and other plastic products and packaging are relatively thin and easy to be blown up by the wind, floating in the air or hanging on branches, scattered in some scenic spots and on both sides of the roads, forming "white pollution".

---

1. Han Lizhao, Wang Tonglin, Yao Yan. Study on the present situation and control countermeasures of "white pollution" [J]. *Population, Resources and Environment of China*, 2010, 20 (S1): 402-404.

It should be pointed out that the plastic waste in the stage of "visual pollution" has not been "irreversible". If people can effectively collect and properly dispose of the plastic waste scattered in the natural environment in time, this "white pollution" will be strangled in the stage of "visual pollution" without further disintegration in the natural system, forming microplastics[1], which will cause deeper harm to the ecology and environment.

2.2.2 Water pollution caused by plastic waste

The water plastic pollution caused by plastic waste refers to plastic waste leaching into rivers, lakes, oceans, due to poor management, affecting the water ecology and environment. The water pollution is mainly divided into land-sourced pollution and sea-sourced pollution. The former is mainly due to plastic waste directly leaking into various water bodies, or in the form of microplastics; sea-sourced pollution mainly refers to fishing and mariculture process of various fishing gear, shipping and marine operations using marine equipment and production of household waste, as well as plastic products carried by tourists are discarded into the ocean after using. Plastic pollution in water is more insidious than plastic pollution in soil, with wider consequences (see Figure 2-2).

These plastic waste in oceans, through the unceasing movement of the ocean and weathering, are always decomposed into microplastics to enter the material cycle, or suspended in the waters and accidentally eaten by marine organisms, therefore becoming part of the biosphere, causing great damage to marine biodiversity. The microorganisms and algae attached to the surface of plastic waste can release an appetizing smell to marine organisms, and its color and shape are similar to those of jellyfish, therefore, lots of maritime animals have eaten them by mistake. It is estimated that about 118 of the 693 species on the International Union for Conservation of Nature's Red List are seriously threatened by plastic pollution.[2]

2.2.3 Soil pollution caused by plastic waste

Compared with "visual pollution", plastic pollution in the soil is rather non-tanglble. Plastic pollution in soils refers to the infiltrating of plastics into the soil in the form of plastic debris or microplastics, resulting in changes in the soil's original properties and state, leading to poor soil permeability or reduced fertility. Soil pollution mainly arises from plastic waste, tire wear particles in road runoff, and agricultural film and pesticide bottles and other agricultural materials used after the random discard. In addition, the use of microplastic-containing animal manure and sludge compost, the use of microplastic-containing sewage for irrigation, etc. will also bring microplastics into the soil (see Figure 2-3).

---

1. Microplastics: plastic fragments and particles smaller than 5mm in diameter.
2. Matthew MacLeod, Hans Peter H. Arp, Mine B. Tekman, Annika Jahnke. The global threat from plastic pollution [J]. *Science*, 2021: 6550.

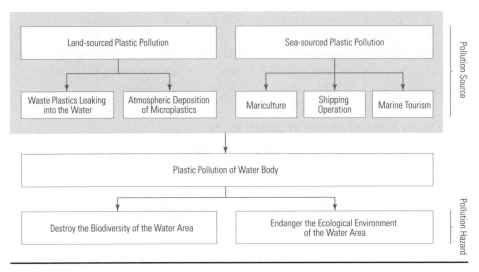

Figure 2-2 Marine Plastic Pollution Path Map

It is estimated that the current plastic content of soil may be more than that in the ocean, and the plastic fraction of soil organic carbon has been as high as 0.1%.[1] A survey of wastewater treatment plants indicates that about 90% of microplastics accumulate in sludge after sewage treatment.

The continuous accumulation of plastic wastes in the soil will not only jeopardies its air permeability, but also hinders the growth of plant roots. What's worse, after microplastics enter the soil, they will absorb a large number of heavy metals and organic pollutants such as pesticides and herbicides in the soil and lock them in the soil environment. This will bring damages to the health of the soil ecosystem, and the activities of soil organisms will also accelerate the spread of microplastics, therefore causing a vicious circle of microplastics spreading in the soil.[2]

---

1. Matthew MacLeod, Hans Peter H. Arp, Mine B. Tekman, Annika Jahnke, The global threat from plastic pollution[J]. *Science*, 2021: 6550.
2. Horton A. A., Walton A., Spurgeon D. J., et al.. Microplastics in freshwater and terrestrial environments: Evaluating the current understanding to identify the knowledge gaps and future research priorities[J]. *Science of the Total Environment*, 2017, 586: 127-141.

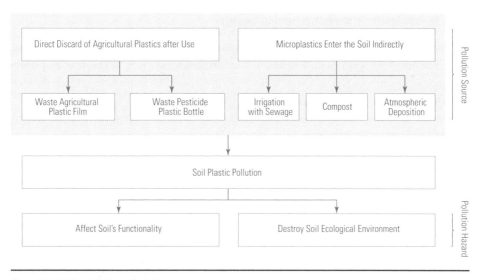

Figure 2-3 Soil Plastic Pollution Path Map

2.2.4 Plastic waste disposal methods affect greenhouse gas emission

Plastics are essentially transformed from chemicals and fossil fuels, namely greenhouse gases are emitted throughout the life cycle of production, distribution, consumption, disposal, and utilization. Some data show that 280-360 million tons of fossil fuels are used to produce plastics worldwide each year[1], and carbon in fossil fuels is transferred to plastics in the process. If these plastic wastes are degraded, incinerated or landfilled, the remaining carbon will eventually be released in the form of carbon dioxide, methane and other greenhouse gases into the air.

---

1. Dees J. P., Ateia M., Sanchez D. L. Microplastics and their degradation products in surface waters: A missing piece of the global carbon cycle puzzle[J]. *ACS ES&T Water*, 2020(1): 214-216.

2.2.5 Microplastics may harm human health

Plastic pollution not only endangers the ecology and environment but also threatens against human health. Microplastics leached into the environment can easily be absorbed by plants and mistakenly eaten by fish and small animals, thus entering the food chain, passing through each step of it and eventually enter the human body (see Figure 2-4). In addition, plastic microbeads are widely applied in toothpaste, body wash and other daily toiletries, while some microplastics can be directly ingested by humans. Some studies even suggest that the ingested microplastics will enter the body's circulatory system and reach specific tissues, potentially causing oxidative stress, inflammatory responses, metabolic disorders, and even affecting the expression of DNA information and genetics.[1]

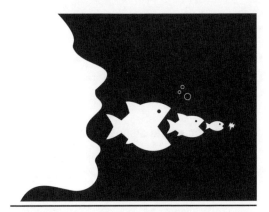

Figure 2-4 Schematic Diagram of Cumulative Conduction of Microplastics
Image: https://baijiahao.baidu.com/s?id=1691641723892993408&wfr=spider&for=pc.

---

1. Landrigan, P. J., Stegeman, J., Fleming, L., Allemand, D., Anderson, D., Backer, L. et al.. Human health and ocean pollution[J]. *Annals of Global Health*, 2020, 1: 1-64.

2.3 Causes of Plastic Pollution

There are a wide variety of synthetic materials widely applied in daily life, but why does plastic pollution attract global attention? To answer this question, we need to analyze from two aspects: the properties of plastic and improper disposal.

2.3.1 Plastics are difficult to resolve

Plastics are difficult to resolve under natural conditions, resulting in cumulative plastic pollution. With a stable physicochemical structure, plastic products are completely assimilated by microorganisms in the natural environment, degrading into CO_2 and water, and its inorganic mineralization may take 200-400 years. According to relevant researches, polystyrene degrades only 0.01%-3%[1] in 4 months in soil, sludge, decaying garbage, or manure microbial communities. The long resolving process of plastic is the root of becoming a "big hazard".

2.3.2 Low value of certain plastic products

Plastics are used not only in the production of major equipment like aircraft and ships but also in packaging and others. Some small food packaging weighs only a few grams, small and light. Recycling these plastic packaging in small sizes is relatively economically costly, but with little in return. Even considering the benefits of resource conservation and environmental protection, it is difficult to maintain such an effective operating mechanism. In addition, the huge variety of plastics products also poses a great challenge for sorting, collection, and recycling. A large amount of plastic waste are consequently leaked into the environment, becoming rampant pollution.

---

1. Yu Y., Yang J., Wu W. M., et al.. Biodegradation and Mineralization of Polystyrene by Plastic-Eating Mealworms: Part 2. Role of Gut Microorganisms[J]. *Environmental Science & Technology*, 2015, 49(20):12087-12093.

2.3.3 The excessive use and inappropriate disposal of plastic waste

(1) Plastic waste leakage into environment caused by improper human behaviors.

Plastic is very common nowadays yet is easy to discard at will, which brings great difficulties to the recycling and disposal of plastic waste, ending with plastic leakage into the environment. It can be said that the improper behavior of humans is the direct cause of plastic pollution.

(2) Disposal and recycling facilities of plastic waste need to be improved.

This is a common issue for most of the countries. Some countries don't have recycling facilities; therefore, their plastic waste still needs to be exported to other countries for recycling. But due to the high cost of long-distance transportation, those low-value plastic waste with little economic value for recycling will be neglected, resulting in an overall low recycling rate. Some countries don't have enough capacity in incineration facilities or standardized landfill facilities, and most of the waste are landfilled randomly.

Developed countries and regions such as the EU and the U.S. face similarly aggravating situation too. In 2018, a total of 29.1 million tons of plastic waste were generated in 30 European countries, including 28 EU members and Norway and Switzerland, of which 12.4 million tons were disposed of by incineration for energy recovery, 9.46 million tons were recycled for materialization and 7.25 million tons were directly landfilled, accounting for 42.6%, 32.5% and 24,9%, respectively.[1] In 2018, the United States generated 35.68 million tons of various types of plastic waste, but only 3.09 million tons were materialized in recycling, accounting for only 8.66%; 5.62 million tons were recycled as energy, accounting for 15.75%; and the volume for landfilling was up to 26.97 million tons, accounting for 75.59%.[2]

---

1. *European Plastics production, demand and waste data Analysis report* (2020).
2. U.S. Environmental Protection Agency, https://www.epa.gov/facts-and-figures-about-materials-waste-and-recycling/plastics-material-specific-data.

(3) Cross-border transfer of plastic waste causes environmental risks.

At present, in the face of increasingly serious plastic pollution, some countries do not assume their responsibility of plastic pollution management, but export the plastic waste collected in their countries to other countries instead, bringing great pressure on other countries and regions to control plastic pollution. According to statistics, the total amount of global plastic waste exported in 2020 was as high as 3.85 million tons, of which the top ten exporting countries exported a total of 2.6 million tons, accounting for 67.5%. The United States has the largest exporting volume of 620,000 tons, followed by the Netherlands, 346,000 tons. The majority of these plastic wastes are exported to Turkey, Malaysia and other places, bringing certain threats to the local ecological environment.[1]

---

1. UN Comtrade Database.

3.

The Concept and Path of Plastic Pollution Control in China

As the largest developing country in the world, China attaches great importance to the treatment of plastic pollution. By vigorously developing a circular economy for plastics and carrying out plastic whole-chain governance, a path on plastic pollution control featured with Chinese characteristics has been developed.

3.1 Historical Evolution of Plastic Pollution Control in China

In ancient times, China established the idea of "the unity of man and nature", advocating nature, respecting nature, and advocating the concept of harmonious symbiosis between man and nature. We began to pay attention to the problem of plastic pollution a long time ago, and took measures to deal with it actively.

(1) Early stage focusing on key sectors of plastic pollution.

After the reform and opening up, with the rapid development of China's economy and the continuous improvement of people's living standards, the consumption of plastics has increased rapidly, and the resulting "white pollution" problem has gradually emerged. China's plastic pollution control at this stage is mainly oriented by outstanding problems. Issued some opinions on strengthening the management of plastic packaging waste in key traffic lines, river basins and tourist scenic spots, "urgent notice on immediately stopping the production of disposable foamed plastic tableware", "Circular of the General Office of the State Council on restricting the production and sale of plastic shopping bags" and other restrictions or prohibit the use of plastic packaging waste, disposable foamed plastic tableware and plastic bags.

(2) Comprehensive solution stage with the development of a circular economy for plastics.

At the end of 1990s, the restriction of resources and environment on economic and social development has become increasingly prominent. China has gradually introduced the concept of circular economy and realized that the treatment of plastic pollution cannot be treated by piecemeal measures, and systematic treatment must be carried out. In 2005, relevant departments of the Chinese government launched a pilot project for the demonstration of circular economy, including the pilot project for the recycling of waste plastics. In 2009, China formally implemented the Circular Economy Promotion Law, comprehensively promoting the construction of a resource recycling system covering the whole society.

(3) A new development stage with whole-chain governance on plastics pollution.

With the rapid development of emerging express, delivery and e-commerce, more and more disposable plastic products are produced, and the plastic pollution is becoming increasingly serious as a consequence. In order to solve this issue, China has comprehensively strengthened the treatment of plastic pollution, issued Opinions on Further Strengthening the Treatment of Plastic Pollution and the Action Plan for the Treatment of Plastic Pollution for the 14th Five-year Plan, and further improved the plastic waste recycling system, so as to promote the treatment of plastic pollution in China into the whole-chain governance. In the practice of plastic pollution control for many years, the legal and policy system of plastic pollution control in China has been continuously improved, the field and scope of coverage have been continuously expanded, and the control efforts have become stronger and stronger, and a closed-loop management system with whole-chain has been gradually formed (see Figure 3-1).

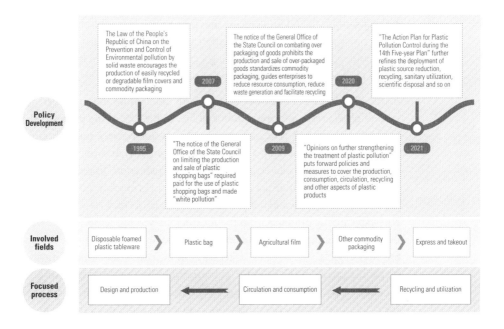

Figure 3-1 History of Plastic Pollution Control in China

3.2 Overall Concept of Plastic Pollution Control in China

As the largest developing country in the world, China has faced more prominent resources and environmental constraints in the process of rapid development over the past few decades and is more determined to take the path of sustainable development. Under this background, China abandons the traditional linear economic growth model of "mass production, mass consumption, and mass waste" characterized by high consumption of resources and energy and high emission of pollutants, and vigorously develops a circular economy. We will explore a sustainable development model in which economic growth is decoupled from resources and the environment.

China is the third country in the world to enact special laws on circular economy. Unlike Germany's circular economy, which focuses on solid waste management, China develops circular economy in production, circulation and consumption, and implements the "3R" principle in all links and the whole process of production, circulation and consumption, that is, " reduce, reuse, and recycle", gradually building a circular economy development model of "resources-products-recycle-resources".

Plastic is an indispensable material. It is both unscientific and unrealistic to remove plastic from our production and life. In the face of the potential risk of plastic pollution, it is necessary to strengthen the recovery and utilization of plastic waste and develop a plastic circular economy. According to the "3R" principle of circular economy, the first is "reduction" to reduce the production and use of disposable plastic products as much as possible, and vigorously promote ecological design methods such as easy recycling and renewable plastic products from the source, followed by "reuse". In the link of circulation and consumption, explore recyclable plastic products and business models. The last part is "recycle". In the end of the disposal process, we should carry out the recovery and material utilization of waste plastic products, carry out energy recovery and utilization of those that do not have the conditions for material utilization temporarily and build a treatment system covering the whole-chain of plastics pollution (see Figure 3-2).

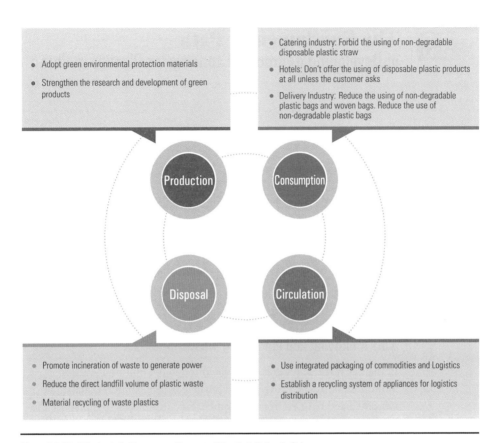

Figure 3-2 The Whole-chain Governance Concept of Plastic Pollution in China

3.3 The Fundamental Path of Developing a Circular Economy for Plastics in China

The development of China's plastics recycling economy has gone through a process of gradual improvement from short-term inhibiting measures to long-term recycling system construction, and from emphasizing end-of-life disposal to whole-chain governance. China has gradually explored a whole life cycle governance system and governance path with Chinese characteristics of plastic pollution, emphasizing effective governance in the whole process of plastic raw material production, ecological design of plastic products, plastic product consumption, plastic waste recycling, and final plastic waste safe disposal (see Figure 3-3).

3.3.1 Improving and innovating plastic materials

In the raw material production stage, through the improvement and innovation of plastic materials, the possible plastic pollution can be prevented and reduced from the source. First, it is necessary to reduce the addition of toxic and harmful substances, such as plasticizer[1], antioxidant[2], anti-adhesion agent[3] and other substances in the process of material production, or the use of high-performance, less toxic resin materials. In addition, it is necessary to improve the performance of plastics, add appropriate modifiers to plastics, improve the service life of plastic products, and effectively improve the recyclability of plastic products after being discarded, such as improving the toughness of agricultural plastic film to improve the recyclability of waste plastic film, so as to reduce the harm of plastic film residue to soil and so on.

---

1. Used to enhance the plasticity.
2. Used to inhibit the oxidation of materials.
3. Used to reduce friction coefficient or static interference.

Figure 3-3 Basic Path Map for the Development of Plastic Circular Economy in China

3.3.2 Promoting eco-design of plastic products

At the design stage, if the existing technical conditions, raw materials and other factors are fully considered to optimize the solution of resource and environmental issues in all aspects, it is possible to maximize resource conservation and reduce environmental pollution from the source.[1] Therefore, in the design of plastic products, integrate the concept of whole life cycle management, carry out systematic evaluation of the social, economic and environmental impacts of plastic products in the production, distribution, consumption and post-waste recycling or disposal, and take measures to improve product design, use more renewable and single plastic products raw materials, and adopt more reusable and easily recyclable product design solutions to achieve the purpose of minimizing the environmental impact of plastic products in the whole life cycle process.

---

1. "Guidance of the Ministry of Environmental Protection of the Development and Reform Commission of the Ministry of Industry and Information Technology on the Development of Ecological Design of Industrial Products", http://www.gov.cn/zwgk/2013-02/27/content_2341028.htm.

3.3.3 Reducing unnecessary consumption of certain single-use plastic products

Due to the low cost and ease of use of disposable plastic products, plastic pollution management has become a "small item, big trouble". In this regard, China has introduced a series of laws and regulations and regulatory requirements, requiring shopping malls, supermarkets and other operational services of a certain area, disposable plastic shopping bag charges; restrict the use of disposable plastic catering utensils in the restaurant industry, to encourage the provision of reusable catering utensils; require hotels and other business premises shall not take the initiative of a certain area to provide disposable toothbrushes, disposable combs, disposable toiletries and other disposable plastic products. And for e-commerce, courier, takeaway and other emerging areas introduced special provisions to encourage e-commerce, takeaway and other platform companies and courier companies to develop and implement plans to reduce the amount of single-use plastic products, and take a variety of measures to encourage consumers to reduce the use of single-use plastic products.

3.3.4 Scientific sorting and recycling of plastic waste

Waste sorting is the key to realize the recycling of plastics. In the process of sorting and recycling, different recycling modes should be adopted according to the product characteristics, circulation characteristics, disposal destination and economic value of the waste (see Figure 3-4). For plastic parts of home appliances, automobiles and other high-value products, it is suggested to take the "accompanied recycling" model, build complete waste recycling system for electrical and electronic products, end-of-life vehicles and other waste products. Regarding to plastic packaging waste with a certain economic value, for example, PET beverage bottles, HDPE daily products packaging barrels (bottles), PP plastic lunch boxes and others. For PET beverage bottles, HDPE daily-use product packaging drums (bottles), PP plastic lunch boxes and others, China has adopted the "special recycling" model and established a large recycling network covering cities and villages, which can basically realize the "collection of all waste plastic bottles". For plastic waste, such as plastic pesticide bottles and mulch, which are widely used, difficult to collect with high risk of environmental leakage, China has adopted a "mandatory recycling" model and has formulated the "Measures for the Management of Agricultural Films" and "Measures for the Management of Pesticide Packaging Waste Recycling and Disposal" to stipulate the recycling obligations of producers. Now, companies have adopted the deposit system to collect and recycle pesticide bottles. For plastic bags, plastic packaging film and other low-value plastics with high recycling costs, it is required to collect and recycle with other household waste.

Plastic Parts	High Value Packages	Pesticide Bottles and Agricultural Films	Low-value Waste Plastic
Build and improve recycling system, strengthen the classification and recovery of plastics, and improve the recovery rate	Engaging individuals and enterprises in the recycling of waste plastic bottles to build a recycling system covering urban and rural areas for a full collection	implement the extension of producer responsibility system to promote agricultural producers and operators Xia Xing pesticide packaging and plastic film recycling obligations-deposit system	Classify with other domestic waste to get unified and clean plastic waste

Figure 3-4 Different Types of Waste Plastics Recycling Mode

3.3.5 Recycling and utilizing plastic waste

At present, plastic waste is mainly recycled in two ways: physical recycling and chemical recycling. Physical recycling[1] is the best choice for the disposal of waste plastic products. This method is simple and feasible and is mainly used to recycle cleaner waste plastics with single composition. Chemical regeneration[2] is mainly for plastic waste that is difficult to be physically recycled, such as plastic film waste, which can realize the recycling of resources, effectively improve the efficiency of plastic waste disposal and reduce the environmental pollution caused by landfills and incineration, and with chemical regeneration technology, companies can manufacture products of the same quality, and this is one way to realize the high-value valorization of plastic waste.

---

1. Physical recycling of plastic waste (also known as physical recycling or physical recovery) refers to the physical processing of pretreated waste plastics into recycled raw materials by physical means such as melting granulation, which is generally divided into melting regeneration and modified recycling.
2. Chemical regeneration of plastic waste (also known as chemical recycling or chemical recovery) refers to the use of chemical technology to convert plastic waste into resin monomer, oligomer, cracking oil or syngas, which can be divided into pyrolysis recovery method and chemical decomposition recovery method.

3.3.6 Processing energy recovery for plastic waste

Plastic waste that cannot be recycled as raw material will enter the domestic waste treatment system. These mixed plastics are difficult to sort, clean, and recycle under the existing technical, and can only be incinerated to generate energy. Through incineration, it can significantly reduce the amount of plastic waste accumulation and landfill land occupation, so that the waste plastic capacity reduction of 90%-95%.[1] But at the same time, the plastic waste in domestic waste also contains a small amount of polyvinyl chloride, polyacrylonitrile, etc., these plastics in the combustion of harmful substances and greenhouse gases, how to do a good job in the incineration process is also critical to control pollutants.

3.3.7 Strengthening the guidance on green consumption and education on eco-friendly lifestyle

The treatment of plastic pollution also calls for extensive participation of the public. China has always strengthened the promotion and guidance of consumer behavior and green education in plastic pollution control. As early as 2007, after the launching of China's plastic pollution control policy, the public has been suggested to use environmentally friendly cloth bags and reduce the using of plastic products. In 2020, after the launching of the new version of the plastic pollution control policy, a variety of green consumption week, DIY environmental protection bags, environmental protection bag design competition and other types of activities are numerous, so that the green concept of environmental protection to spread. In addition, to cultivate the successor of green consumption, ecological and environmental protection propagandists and practitioners, many schools to carry out "Plastic pollution control" publicity activities, so that the concept of green consumption could be passed on from generation to generation.

---

1. Hou Caixia. *Study on degradation of plastics by supercritical water* [D]. Tianjin University, 2003.

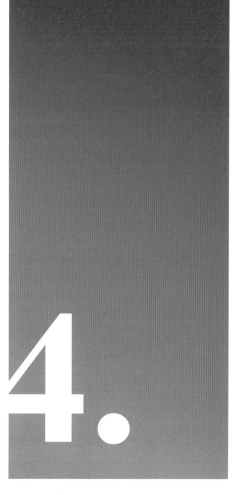

4.

Plastic Pollution Control System and Accomplishments in China

After decades of efforts, China has established a relatively sound plastic recycling system through the development of a circular economy and the control of plastic pollution from its whole life cycle, with a great many accomplishments.

4.1 Plastic Pollution Control System in China

The Chinese government has introduced a series of incentive measures such as investment, finance and taxation to standardize and guide the recycling of plastic waste and promote the green, low-carbon and recycling development of the plastics industry. With the promotion of the government, enterprises and wide public participation, a wide range of recycling systems have been formed.

4.1.1 Formulating laws and regulations on plastic pollution control

(1) Incorprating the plastic pollution prevention and control into the fundamental law of the environment.

The Environmental Protection Law of the People's Republic of China and the Law of the People's Republic of China on the Prevention and Control of Environmental pollution by solid waste include the treatment of plastic pollution from the point of view of protecting the water environment, prohibiting the active provision of disposable plastic products, and encouraging the optimization of packaging.

(2) Promulgating departmental regulations and rules on plastic pollution control departments for key sectors.

The Chinese government has issued Measures for the Management of Packaging for the Delivery Industry, Measures for the Management of Agricultural Film and Measures for the Recycling and Disposal of Pesticide Packaging Wastes in the agricultural sector, and Regulations on Pollution Prevention and Control of Waste Plastics Processing and Utilization in the industrial sector.

4.1.2 Improving standard system on the plastic pollution control constantly

(1) Improving the standards on the design and production of plastic products.

The Chinese government have formulated a series of national standards for ecological design, such as "General Principles for Evaluation of Ecological Design Products", "General Principles for Evaluation of Electronic and Electrical Ecological Design Products", "General Principles for Ecological Design Product Identification" and "General Principles for Design and Evaluation of Easy Recycling and Renewable Plastic Products", in order to improve the recovery rate of plastics.

(2) Improving the standard on the recycling and utilization of plastic waste.

Chinese government have formulated the Technical Specification for Recycling and Utilization of Waste Plastics, part 1 of recycled Plastics: general principles, and Technical Specifications for Pollution Control of Recovery and Recycling of Waste Plastics (for trial implementation). Specific requirements are put forward for environmental protection-related matters in the process of recycling and recycling of waste plastics.

4.1.3 Introducing incentive policies and measures to control plastic pollution

(1) Formulating financial policies on plastic recycling.

The Chinese government launched financial subsidies for the construction of plastic waste recycling projects and domestic waste collection and incineration facilities and subsidizes the price of electricity generated from the incineration of plastic waste and other domestic waste. The government has introduced the "Guidelines for Issuing Green Bonds" and "Green Credit Guidelines" to give priority to relevant environmental projects for financial support such as issuing bonds.

(2) Formulating preferential tax policies on plastic recycling.

The Chinese government has issued the Catalogue of Preferential Value-Added Tax for Products and Services for Comprehensive Utilization of Resources, granting a larger scale of preferential policy of immediate VAT refund for recycled plastics and other related products; and the Catalogue of Preferential Income Tax for Enterprises for Comprehensive Utilization of Resources, reducing the amount of income tax paid by enterprises in this industry.

4.1.4 Formulating the joint mechanism among the government, enterprises, and the public

(1) Government departments are responsible for overall planning and developing infrastructures.

Chinese government has formulated documents such as Opinions on Establishing a Complete and Advanced Recycling System for Used Goods and Guiding Opinions on Accelerating the Development of the Renewable Resources Industry to promote the construction of waste recycling systems. Under the guidance of the national and local governments at all levels, a relatively complete recycling system has been constructed with an integrated urban and rural waste collection and treatment system of "waste collected across villages, transferred through towns and treated in county level". The system can help with domestic waste removal and disposal, effectively preventing plastic waste from leaking.

(2) Enterprises are responsible for the whole life cycle control of plastic pollution.

Enterprises shall follow national regulations and rules in carrying out R&D and the innovation of alternative products and packaging in manufacturing plastic products. In circulation phase, enterprises shall stop offering disposable plastic shopping bags, etc. In the recycling phrase, enterprises are encouraged to build plastic waste recycling facilities, invest more on the R&D of recycled plastic production technology to reduce the risk of environmental leakage of plastic waste.

(3) The public are encouraged to participate in green consumption and garbage segregation.

The public is the practitioner of plastic pollution control, and more and more consumers voluntarily choose environmentally friendly products, bring their own toiletries when traveling, and refuse to over-package goods. The new trend of green consumption has therefore been gradually formed. In daily life, the generally public voluntarily sorts the garbage, litter plastic waste into designated bins , and give recyclable plastic waste to specialized recyclers, helping to recycle plastic waste.

4.2 Accomplishments of Plastic Pollution Control in China

4.2.1 China has built the large-scale plastic waste recycling system with extensive coverage

China has established a comprehensive waste plastics recycling system consisting of recycling outlets, sorting centers and processing and utilizing plants, and has made use of the Internet, the Internet of things and other technological innovations in recycling models to promote the integration between the waste separation network and the recycling network. Remarkable results have been achieved. At present, China has the largest waste plastics recycling capacity in the world, and the number of enterprises engaged in waste plastics recycling and reclamation exceeds 15,000, with about 900,000[1] employees. In recent years, the average annual growth rate of waste plastics recycling in China has been remaining at a consistent 2.5%. In 2021, China produced about 62 million tons of waste plastics, of which the materialized recycling volume was about 19 million tons, 31% of that has been recycled as materials, which is nearly 1.74[2] times of the global average recycling rate (see Figure 4-1). China's recycling capacity in this regard accounts for about 70% of the world. In 2018, the recycling rate of local plastic waste in the U.S. was only 5.31% and that in EU was only 17.18%. The number of Japan in 2019 was 12.50%[3] with the total amount of materialized waste was 7.78 million tons, and China's total materialized waste was 2.43 times of that in 2019.

---

1. *Report on Employees in Recycled Plastics Industry* [R]. China National Resources Recycling Association, 2022.
2. *Report of Plastic Recycling Industry in China (2020-2021)* [R]. Recycled Plastics Association, China National Resources Recycling Association, 2021.
3. *Report of Plastic Recycling Industry in China (2020-2021)* [R]. Recycled Plastics Association, China National Resources Recycling Association, 2021. *Data Analysis Report on European Plastic Production*, Demand and Waste (2020) / Plastics Europe. Ministry of Environment website (epa.gov) / U.S. Bureau of Statistics U.S. Census Bureau. Plastic Products, Waste and Resource Recycling in Japan 2019, Japan Plastics Recycling Association (PWMI).

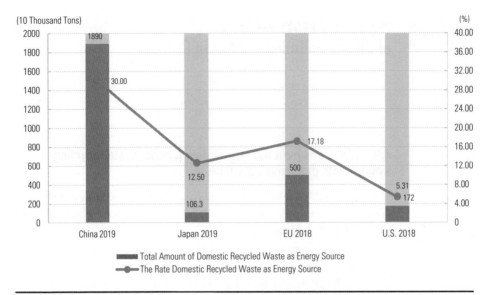

Figure 4-1 Comparison of Domestic Material Recovery between China and Other Countries
Source: *Report of Plastic Recycling Industry in China (2020-2021)* [R]. Recycled Plastics Association, China National Resources Recycling Association, 2021.Data Analysis Report on European Plastic Production, Demand and Waste (2020) / Plastics Europe. Ministry of Environment website (epa.gov) / U.S. Bureau of Statistics U.S. Census Bureau. Plastic Products, Waste and Resource Recycling in Japan 2019, Japan Plastics Recycling Association (PWMI).

4.2.2 A comprehensive recycling system for plastic waste has been established

China has established a comprehensive waste plastic recycling system, covering from low-value products such as flowerpots and garbage cans to high-value products such as home appliances and automobiles, so that recycled plastics are widely used in textiles, automobiles, packaging and many other areas. Different types of plastic wastes are used in different ways: firstly, we use physical regeneration methods to process plastic wastes into recycled plastics with the same or similar properties as virgin plastics for product production, such as the "bottle to bottle" utilization of waste beverage bottles. Secondly, we use certain processing methods for some waste plastic products that are heavily polluted and difficult to clean. Plastic products, using certain processing methods, are processed into recycled plastics with slightly lower performance than virgin plastics, which are used to produce some relatively low-end products, such as flowerpots and garbage cans. Thirdly, the chemical components in the recycled plastic waste are extracted to make them into monomers or fuels. Fourthly, the energy in waste plastics is directly used for incineration and power generation through waste incineration (see Figure 4-2).

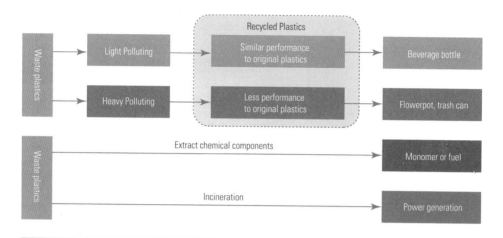

Figure 4-2 Differentiated Recycling of Waste Plastics

Since 2016, China's solid waste incineration facilities in municipal levels have increased year by year, with the increasing daily processing capacity. As shown in Figure 4-3, the number of municipal solid waste incineration facilities has increased from 249 in 2016 to 463 in 2020, nearly doubled. The incineration capacity increased from 255,850 tons/day to 567,804 tons/day, increased by 1.2 times. The improvement of processing capacity has significantly increased the incineration treatment capacity of municipal solid waste. In the past five years, the incineration treatment capacity has increased from 73.78 million tons per year to 146.08 million tons per year, with a compound annual growth rate of about 30%. In 2020, the proportion of municipal solid waste incineration has reached 62.29% (see Figure 4-3). According to China Plastic Recycling Association's calculations, among them, the annual energy utilization of plastic waste reached 27.4 million tons, and the energy utilization rate was 45.7%.

With the continuous improvement of China's plastic recycling industry, the output value of waste plastic recycling is also increasing. In 2019, the output value of waste plastics recycling in China reached 100 billion yuan; in 2021, that number reached 105 billion yuan, an increase of 33%[1] over the same period last year.

---

1. *Report of Plastic Recycling Industry in China (2020-2021)* [R]. Recycled Plastics Association, China National Resources Recycling Association, 2021.

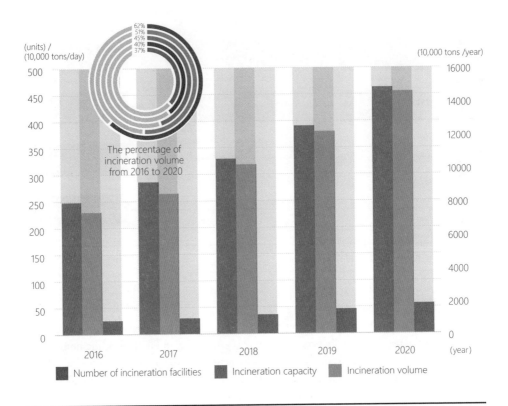

Figure 4-3 Statistics of Waste Incineration Facilities in China in 2016-2020 (Counties Not Included)
Note: The percentage of incineration volume in the calendar year is calculated according to the proportion of the actual treatment volume.
Source: Ministry of Housing and Urban-Rural Development. *Urban Construction Statistical Yearbook (2016-2020)*.

4.2.3 The utilization of recycled plastics effectively reduces the consumption of fossil raw materials

Plastic waste could be resources and also pollutants. If effectively recycled, it could be the recycled resource. If not handled properly, it will become pollutants. Therefore, we need to recycle the plastic waste in a practical way, "turning waste into treasure". This can transform waste into resources, and made great contribution to the economic and social development, thereby reducing the excessive consumption of non-renewable natural resources such as oil by humans.

From 2016 to 2020, China has recycled a total of 108 million tons of waste plastics, with a total value of more than 600 billion yuan. If we calculate that recycling 1 ton of waste plastics is equivalent to saving 3 tons of oil, China has saved 330 million tons of oil extraction and

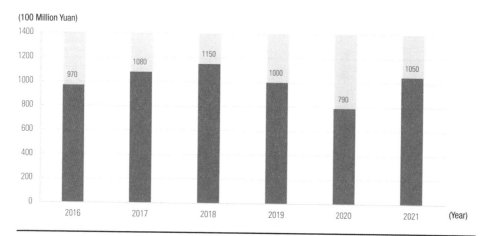

Figure 4-4 Recycling Value of Waste Plastics in China from 2016 to 2021
Source: Recycled Plastics Division of China National Resource Recycling Association, The Ministry of Commerce of the People's Republic of China.

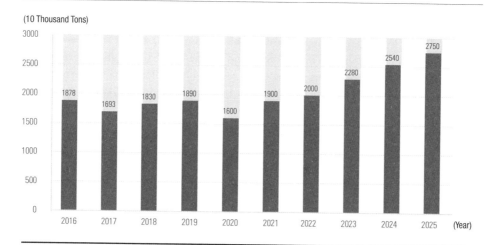

Figure 4-5 2016-2025 Waste Plastics Recycling Quantity in China
Source: China National Resource Recycling Association.

consumption[1]. At the same time, China also treated a large amount of plastic waste from other countries. Since 1992, China has imported and recycled 106 million tons of waste plastics[2], saving the world 318 million tons of crude oil extraction and consumption (see Figure 4-4 and Figure 4-5).

---

1. The Ministry of Commerce of the People's Republic of China.
2. The public data of major countries in the world and customs statistics of the People's Republic of China.

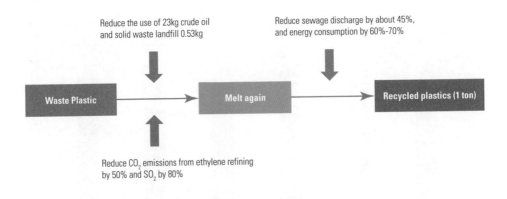

Figure 4-6 Emission Reduction in the Production of Recycled Plastics
Source: Zou Qizhi. Study on emission reduction measures of waste paper and papermaking enterprises[J]. *Resource Conservation and Environmental Protection*, 2014, 6: 33-34.

4.2.4 Reducing pollutants and carbon dioxide emissions

The recycling of waste plastics can reduce crude oil consumption, solid waste generation, and emissions of pollutants such as CO_2, SO_2, and sewage. According to the relevant ratio conversion (see Figure 4-6), China has recycled a total of 170 million tons of waste plastics of all kinds from 2011 to 2020, which is equivalent to a total reduction of 510 million tons of crude oil consumption, 0.9 billion tons of solid waste emission and 61.2 million tons of CO_2 emission.

5.

Experience from Plastic Pollution Control in China

China has gradually found an effective way to solve plastic pollution through the development of plastic circular economy. Its practice and experience have important inspiration and reference value for further improving and strengthening plastic pollution control and promoting global cooperation in plastic pollution control.

5.1 It is Necessary to Establish a Sound Life Cycle Management System to Control Plastic Pollution

The treatment of plastic pollution is a complex system engineering, which involves multiple procedures such as product design, production, circulation, consumption, collection, treatment, recycling and so on.

(1) It needs to be recognized that the reduction and replacement of disposable plastic products is only one part of life cycle management.

At present, the laws and regulations on disposable plastic products issued by most of countries are beneficial and necessary in solving the plastic pollution. However, it must be recognized that the proportion of disposable plastic products in the whole plastic industry system is relatively low, and its reduction and replacement is only a part of the whole life cycle treatment of plastic pollution, and its contribution to plastic pollution control is limited.

(2) A comprehensive infrastructure for end collection and disposal of plastic waste is the key to the prevention and control of plastic pollution.

The treatment of plastic pollution needs to establish a comprehensive management system from the design and production of plastic products to end disposal and utilization, but the construction of effective collection and disposal facilities of plastic waste can directly prevent the plastic waste leakage. Therefore, countries should give priority to the construction of plastic waste collection and disposal facilities in the treatment.

To Develop Plastic Recycling Economy Calls
for Home-based Recycling System

There are many kinds of waste plastics and various ways of utilization and disposal, so it is necessary to establish a perfect recycling system for all kinds of waste plastics.

(1) Plastic waste recycling system calls for continuous improvement.

We need to gradually improve the traditional recycling model based on private recycling, promote standardized recycling led by corporates, and take good advantage of new technologies such as the internet, intelligent recycling machines, intelligent sorting new equipment to try new recycling models and improve recycling efficiency, reduce recycling costs, strengthen the flow of waste plastics management, and improve the modernization of the recycling system.

(2) Recycled plastics grading is the key to improving recycling rates.

With proper technologies and financial support, conducting different grades of waste plastics recycling system will not only realize a full utilization of specific species, but also realize the differentiated utilization of waste plastics in wood plastic, express packages and other products for the best end application. In the way, the recycling rate of plastic products across the society could be effectively improved.

(3) The sorting of plastic wastes should base on its end disposal and utilization.

Plastics contained in consumers electronics, automobiles and other products should be separated before recycling. For the separated and individual plastic products such as beverage bottles and plastic lunch boxes, they should be taken as recyclables, and a perfect recycling system should be established to realize the classified recycling of these waste plastic products. For the hybrid plastics that are seriously polluted and difficult to sort, energy utilization is still a better way, and there is no need for classified recycling.

5.3 The Development of Circular Economy for Plastic
Requires Comprehensive Consideration of Environmental and Economic Benefits

When developing circular economy for plastic, only considering environmental and economic benefits can we achieve sustainable development.

(1) It is necessary to establish a sustainable development model of circular economy for plastics.

In developing plastic circular economy, the economic benefits can help to stimulate stakeholders. The development model that only pursues environmental benefits while ignoring economic benefits is not sustainable. However, at the same time, it is necessary to establish a value compensation mechanism through levying funds and giving subsidies to ensure that the recycling behavior of stakeholders is economic and encourage the recycling of waste plastics with little economic value.

(2) Technological innovation is the key to waste plastics recycling.

Whether the waste plastics can be recycled or not depends to a large extent on the breakthrough of recycling technology, including technology and equipment for high-value utilization of recycled plastics, development and application of green modifiers, pollution prevention and control in the process of waste plastic regeneration, energy utilization and pollution prevention of low-value waste plastics can significantly improve waste plastics recycling, and achieve the balance between economic and environmental benefits to promote the development of plastic recycling economy.

(3) Business model innovation helps to achieve a balance between economic and environmental benefit.

Innovative recycling models can effectively promote the recycling of waste plastics and reduce recycling costs. For example, by implementing a deposit system to encourage end consumers to hand over waste pesticide bottles and beverage bottles to collection points, the cost of collection in the recycling process can be reduced. By adopting the "Internet + Recycling" method to recycle waste household appliances and make online appointments, we can optimize the recycling route, improve the efficiency of recycling, and help reduce the cost as well.

5.4 Developing Circular Economy for Plastic
Requires Scientific Comparative Analysis of Various Plastic Substitute Products and Schemes

Every new scheme and technology needs its life cycle analysis to ensure that it will generate new pollution in the process of plastic pollution treatment.

(1) Chemical recovery still needs to further analysis on its sustainable business models and product solutions.

Under current conditions, chemical recycling is more costly than physical recycling. Therefore, physical recycling is still the best choice for those categories that are suitable. For other categories that physical recycling is not an option, chemical recycling technology should be carefully carried out after the classification of plastic waste by material at the source and to explore a reasonable and feasible commercial operation model. In addition, the comprehensive economic and environmental benefits of plastic waste recycling technology routes that simply use fuel oil as a product still needs scientific evaluation.

(2) Alternative products still need scientific comparative analysis.

At present, to reduce the use of disposable plastics in consumption, there are many alternative products made of paper, bamboo, and other materials. However, there are still space to improve for alternatives in many aspects such as technical feasibility, economics, and scalability. Compared with the use of plastic products, there is a need for a scientific comparison and analysis of various alternatives and programs, both in terms of economics and life-cycle environmental impact, including post-waste disposal.

5.5 The Development of a Circular Economy for Plastics Calls for Cooperation between Government and Enterprises and a Holistic Participation

It is necessary to gather the government and enterprises to develop plastic circular economy and carry out plastic pollution control.

(1) The government should be the regulator and promoter of a circular economy for plastics.

The government needs to promulgate relevant laws and regulations on the treatment of plastic pollution, formulate product ecological design standards, strengthen the construction of long-term mechanisms such as the extension of producer responsibility, and introduce relevant supporting policy systems. and take practical measures to encourage and support the construction of relevant projects to provide necessary conditions for the development of plastic circular economy.

(2) Enterprises should take the initiative in developing the circular economy of plastics.

Plastic products manufacturers should take the initiative to reduce the use of plastic microbeads and bear in mind the principle of "easy to recycle, easy to regenerate" in design. For plastic products using enterprises, it is necessary to develop disposable plastic products and reduce the using of alternatives, while increasing the use of environmentally friendly products. Plastic recycling enterprises should accelerate the construction of a comprehensive waste plastic recycling system, innovative recycling technology and mode.

(3) The general public should be engaged in the circular economy for plastics.

Consumers should take the initiative to reduce the use of disposable plastic products and actively participate in garbage sorting. In addition, the vast number of social organizations and news media should become propagandists of plastic circular economy and popularization of related knowledge, so as to improve the scientific understanding of plastic problems from all walks of life.

5.6 The Plastic Pollution Control Calls for Extensive International Cooperation

No country can be left aside in the face of plastic pollution, and it is necessary to strengthen international exchanges and cooperation to help less developed countries and regions establish a plastic circular economy system as soon as possible.

(1) All countries shall establish their own plastic recycling and disposal system as soon as possible.

Building the holistic recycling system based on domestic recycling model can help reduce recycling costs and improve the recycling rate of waste plastics. Therefore, countries, especially developed countries and regions, shall gradually change the traditional practice of simply collecting plastic waste and exporting it to other countries and regions, and instead realize local recycling and scientific disposal of plastic waste within the country or region.

(2) It is necessary to pay close attention to the underdeveloped countries and regions with weak plastic recycling infrastructure.

Some less developed countries and regions don't have sound plastic waste collection and treatment facilities, and the plastic recycling rate is relatively low, the risk of environmental leakage is high, becoming the weak part of global plastic pollution management. Therefore, "plastic waste" should be prohibited to these countries and regions that do not have better recycling conditions for export. At the same time, developed countries and international organizations should give them appropriate help in terms of funding, technology, management and human resources.

(3) Build a global monitoring and evaluation system for the flow of plastic waste.

We should speed up the construction of a global monitoring and evaluation system for the cross-regional flow of plastic waste, and with the help of material flow analysis method, quantitatively and dynamically track the international flow of plastic waste in order to carry out effective evaluation of the disposal and utilization of plastic waste in various countries and predict future development trends and solve problems as timely as possible.

6.
Initiatives on Enhancing Global Control of Plastic Pollution Control

At present, countries are facing increasingly serious plastic pollution problem, especially the marine plastic pollution and it has increasingly become the focus of the world's environmental protection, bringing great challenges to the sustainable development of mankind.

Report of the National Think Tank
Plastic Pollution Prevention and Control in China: Principles and Practice (Extracted version)

According to a report released by the United Nations Environment Programme in 2021, of the approximately 9.2 billion tons of plastic produced globally between 1950 and 2017, about 7 billion tons became plastic waste, and the recycling rate of plastic was less than 10%. The annual production of plastic waste worldwide is about 300 million[1] tons, and a large amount of plastic waste leak into the soil and the sea, eventually causing "white pollution" and posing a serious threat to ecological protection and biodiversity.

Plastic pollution is a common challenge facing mankind, and no country can be left alone. Therefore, it is necessary to establish a sense of a community with a shared future for mankind, and the whole world should unite to take actions on it, so as to build a "plastic pollution control community" with the extensive participation of countries.

To this end, we'd like to propose the following initiatives:

(1) We believe that the essence of plastic pollution is the leakage into environment. Plastic itself is not a pollutant, and the focus of plastic pollution control is the recycling and disposal. But at the same time, we also realize that the whole life cycle management of plastics pollution is an important way to reduce the pressure of plastic end disposal.

(2) We have observed that the global plastic pollution has been accumulating for many years and transferred across regions. Therefore, countries need to take immediate action not only to effectively control the current plastic waste, but also to take effective measures to control the historical plastic waste.

(3) We believe that global plastic pollution control requires all countries to take immediate action to enhance their facilities on plastic waste collection, disposal and recycling to prevent plastic waste from leaking into the environment.

(4) We believe that in developing plastic recycling economy, we should give priority to recycling waste plastics into raw materials and then energies, and finally the standardized landfill disposal. Random disposal of plastic waste should be strictly prohibited.

(5) We believe that plastic pollution and control require extensive international cooperation and encourage all countries and regions, including the private sector, to strengthen regional, national and local bilateral and multilateral cooperation among all stakeholders.

(6) We believe that countries should exercise reasonable control over the cross-border transfer of plastic waste, and exporting countries should ensure that the receiving countries of plastic waste have sound disposal infrastructure and conditions to avoid

---

1. United Nations (UN) Environment. *Beat Plastic Pollution*[EB/OL]. https://www.unenvironment.org/interactive/beat-plastic-pollution/.

secondary pollution, and provide support or assistance to the receiving countries when necessary.

(7) We advocate that each country and region should formulate and introduce special laws and regulations and action plans for plastic pollution control according to their national conditions, put forward the milestones of plastic pollution control, and take effective actions for it.

(8) We believe that to effectively deal with global plastic pollution control, developed countries and regions need to give necessary financial, technological and human resources support to less developed countries and regions in terms of infrastructure construction and management capacity enhancement.

(9) We believe that plastic pollution, especially marine plastic pollution, requires the establishment of a scientific monitoring and assessment system to make scientific judgments on the formation and cross-regional flow of plastic pollution and to guide the scientific implementation of global plastic pollution control.

(10) We believe that the global plastic pollution control needs to summarize the successful experiences and practices of various countries in a timely manner, form case studies and action guidelines for reference of each country and region and improve the comprehensive capacity of each country and region in plastic pollution control.

Conclusion

At present, plastic pollution has become a global environmental problem second only to climate change, and has attracted great attention from various countries and regions. After decades of continuous exploration and efforts, China has explored a "Chinese solution" to plastic pollution management through the development of plastic recycling economy, making an important contribution to global plastic pollution management. However, we should also see that plastic pollution control is a complex system engineering. There still are many shortcomings and needs to be improved and enhanced. It needs more countries and regions to join, and plastic pollution control still has a long way to go.

However, we also firmly believe that as long as all countries in the world cooperate and actively participate, the plastic pollution problem will be effectively solved in the near future, and the vision of sustainable development in which people and nature live in harmony will be realized!

References

[1] Dees J. P., Ateia M., Sanchez D. L. Microplastics and their degradation products in surface waters: A missing piece of the global carbon cycle puzzle[J]. *ACS ES&T Water*, 2020, 1: 214-216.

[2] *From Pollution to Solution: A Global Assessment of Marine Litter and Plastic Pollution* [R]. United Nations Environment Programme, 2021.

[3] Han Lizhao, Wang Tonglin, Yao Yan. Study on the present situation and control countermeasures of "white pollution" [J]. *Population, Resources and Environment of China*, 2010 20 (S1): 402-404.

[4] Horton A. A., Walton A., Spurgeon D. J., et al.. Microplastics in freshwater and terrestrial environments: Evaluating the current understanding to identify the knowledge gaps and future research priorities[J]. *Science of the Total Environment*, 2017, 586: 127-141.

[5] Hou Caixia. *Study on Degradation of Plastics by Supercritical Water* [D]. Tianjin University, 2003.

[6] Landrigan P. J., Stegeman J., Fleming L., Allemand D., Anderson D., Backer L., et al.. Human health and ocean pollution[J]. *Annals of Global Health*, 2020, 1: 1-64.

[7] Lau W., Shiran Y., Bailey R. M., et al.. Evaluating scenarios toward zero plastic pollution[J]. *Science* (New York, N. Y.), 2020, 6510(369): 1455-1461.

[8] Ma Zhanfeng, Jiang Wanjun. China Plastics processing Industry (2020) [J]. *China Plastics*, 2021 and 35 (5): 119-125.

[9] Macleod M., Arp H., Tekman M. B., et al.. The global threat from plastic pollution[J]. *Science* (New York, N. Y.), 2021, 6550(373):61-65.

[10] Ministry of Industry and Information Technology, Development and Reform Commission and Ministry of Environmental Protection. *Guidance on Developing Ecological Design of Industrial Products* [R]. Department of Energy Saving and Comprehensive Utilization, Ministry of Industry and Information Technology, 2013.

[11] *Plastics-the Facts 2020: An Analysis of European Plastics Production, Demand and Waste Data*[R]. Plastics Europe Association of Plastics Manufactures, 2020.

[12] *Recycling of plastic products, wastes and resources in Japan 2019* [R]. Japan Plastics Recycling Association

[13] *Report of Plastic Recycling Industry in China (2019-2020)* [R]. Recycled Plastics Association, China National Resources Recycling Association, 2020.

[14] *Report of Plastic Recycling Industry in China (2020-2021)* [R]. Recycled Plastics Association, China National Resources Recycling Association, 2021.

[15] *Report on Employees in Recycled Plastics Industry* [R]. China National Resources Recycling Association, 2022.

[16] United Nations (UN) Environment. *Beat Plastic Pollution*[EB/OL]. https://www.unenvironment.org/interactive/beat-plastic-pollution/.

[17] Wangliming. *Development of recycling and utilization technology of waste household appliance plastics* [EB/OL]. https://zhuanlan.zhihu.com/p/354754130, 2021/3/5.

[18] Yu Y., Yang J., Wu W. M., et al.. Biodegradation and Mineralization of Polystyrene by Plastic-Eating Mealworms: Part 2. Role of Gut Microorganisms[J]. *Environmental Science & Technology*, 2015, 49(20):12087-12093.

[19] Zheng Qiang. *Discussion on Plastics and "White pollution"* [EB/OL]. https://mp.weixin.qq.com/s/qLPyAVI2xt49-QiUCOHdPA,2021/11/19.

Appendix

Excerpts from the Summary of Chinese Regulations and Policies Regarding to Plastic Pollution Control over the Years

Type	Formulated/ Revised Time	Policy Document	Main Content
Laws and Regulations	Formulated in December 1989. Revised in April 2014	Environmental Protection Law of the People's Republic of China	The State Council and local people's governments at all levels in coastal areas shall strengthen the protection of the marine environment. The discharge of pollutants into the ocean, dumping of wastes, and construction of coastal and marine engineering shall comply with laws, regulations and relevant standards to prevent and reduce pollution and damage to the marine environment. The state encourages and guides citizens, legal persons and organizations to use products that are conducive to protecting the environment and recycled products to reduce the generation of waste. Local people's governments at all levels shall take measures to organize the classified disposal and recycling of domestic waste
	Formulated in October 1995. Revised in April 2020	Law of the People's Republic of China on the Prevention and Control of Environmental Pollution by Solid Waste	Any entity is prohibited from dumping, stacking and storing solid wastes on rivers, lakes, canals, channels, reservoirs and their floodlands and bank slopes below the highest water level, as well as other locations specified by laws and regulations. The local people's government at or above the county level shall speed up the establishment of a household waste management system for classified release, classified collection, classified transportation, and classified treatment, so as to realize the effective coverage of the domestic waste classification system. Any entity shall dump domestic waste at designated places in accordance with the regulation. It is forbidden to dump, scatter, pile up or burn domestic waste at will. Entities that produce agricultural solid waste such as straw, waste agricultural film and pesticide packaging wastes shall take recycling and other measures to prevent environmental pollution. E-commerce, express delivery, food takeaway and other industries should prioritize reusable and easy-to-recycle packaging materials, optimize the packaging, reduce the use of packaging materials, and voluntarily recycle the packaging materials. According to the law, the state prohibits and restricts the production, sale and use of disposable plastic products such as non-degradable plastic bags. The state encourages reducing the use of disposable plastic products and promotes the recycling of disposable plastic products such as plastic bags. The use of recyclable, easily recyclable and degradable alternatives are also encouraged. Entities in tourism and hotels should not offer disposable products in accordance with relevant state regulations

continued

Type	Formulated/Revised Time	Policy Document	Main Content
Laws and Regulations	August 2008	Circular Economy Promotion Law of the People's Republic of China	Enterprises producing products or packages listed in the catalogue of articles subject to compulsory recycle must be responsible for recycling deserted products or packages.For those usable,the producers thereof shall be responsible for using them,while for those products which are inappropriate for reutilization due to the absence of technical or economic conditions,the producers shall make them harmless.
			For products or packages listed in the catalogue of articles subject to compulsory recycle, consumers shall deliver the deserted ones to the producers or the distributors or other organizations entrusted by the producers for recycle.
			Enterprises engaging in the design of products,equipment,products and packages shall, in accordance with the requirement of reducing the consumption of resources and the generation of wastes, give preference to the materials which are recyclable, dismountable, degradable, innocuous, harmless or slightly harmful or poisonous, and the compulsory requirements in the relevant state standards shall be satisfied.
			Enterprises in the catering, entertainment, hotel and other service industries shall use energy-saving, water-saving, material-saving and environment-friendly products and reduce or stop using energy-waste or contaminating products.
			The state sets restrictions on the production and distribution of one-off consumption goods under the precondition of safeguarding product security and sanitation.The specific directory of the one-off consumption goods under restriction shall be formulated by the administrative department of circular economy development under the State Council together with the public finance department and the competent department of ecology and environment and other relevant competent departments under the State Council.
			The state encourages and advocates the construction of a waste recovery system. The local people's governments shall, according to the urban and rural planning, reasonably position the waste recycling outlets and trading markets, and support waste recycling enterprises and other organizations in the collection, storage, transport and information exchange of wastes
	April 2020	Measures for the Administration of Agricultural Film	It is prohibited to produce, sell or use agricultural films that are explicitly prohibited by the state or do not meet mandatory national standards. Encourage and support the production and use of fully biodegradable agricultural films.
			Agricultural film users shall pick up the non-biodegradable agricultural film wastes in the field before the expiration of the term of use and hand them over to recycling outlets or recycling workers, and shall not discard, bury or burn them at will.
			Agricultural film producers, sellers, recycling outlets, waste agricultural film recycling enterprises or other organizations shall cooperate and adopt various ways to establish and improve the recycling system of agricultural film, promote the recovery, treatment and reuse of waste agricultural film

continued

Type	Formulated/Revised Time	Policy Document	Main Content
Laws and Regulations	August 2020	Administrative Measures for the Recovery and Disposal of Pesticide Packaging Wastes	Pesticide producers (including enterprises exporting pesticides to China), operators and users should voluntarily fulfill their obligations to recycle and dispose of pesticide packaging waste, and promptly recycle and dispose of pesticide packaging waste. Pesticide producers and operators shall fulfill the corresponding obligations of recycling pesticide packaging wastes in accordance with the principle of "whoever produces and operates shall recycle". Pesticide operators shall set up pesticide packaging waste recycling devices at their business premises, and shall not reject the packaging wastes they sell pesticides. Pesticide users should collect pesticide packaging waste in a timely manner and return it to pesticide operators or pesticide packaging waste recycling stations (points), and must not discard them at will. The State encourages and supports the utilization of pesticide packaging waste as resources; other than the utilization of resources, it shall be disposed of in a harmless manner such as landfill and incineration in accordance with laws and regulations. The pesticide packaging waste disposal costs shall be borne by the corresponding pesticide producers and operators; if the pesticide producers and operators are unclear, the disposal costs shall be paid by the local county-level people's government
	November 1990	Measures for the Hygienic Administration of Plastic Products and Raw Materials for Food	Recycled plastics shall not be used for processing plastic tableware, containers and food packaging materials
	December 2005	Guiding Catalogue for Industrial Structure Adjustment (2005 Edition)	It is encouraged to use composite materials and polymer materials, eliminate foam and disposable foamed plastic tableware

continued

Type	Formulated/Revised Time	Policy Document	Main Content
Laws and Regulations	August 2012	Regulations on the Prevention and Control of Pollution in the Processing and Utilization of Waste Plastics	It is forbidden to use waste plastics to produce ultra-thin plastic shopping bags with a thickness of less than 0.025mm and ultra-thin plastic bags with a thickness of less than 0.015mm. Waste plastic processing entities shall dispose of the residual garbage and filters generated during the processing and utilization of waste plastics in an environmental friendly way; it is prohibited to hand them over to entities that do not meet environmental protection requirements for disposal. It is forbidden to incinerate waste plastics in the open air and the residual garbage and filter screen generated during processing and utilization. The waste plastic processing and utilization distribution center shall establish a centralized recycling and disposal mechanism for the residual garbage and filter screens generated by the waste plastic processing and utilization retail households. Encourage waste plastics processing and utilization distribution centers to implement centralized park management of waste plastics processing and utilization retail households, and centrally treat waste water, waste gas and solid waste generated by waste plastics processing and utilization
Normative Documents	September 1989	Opinions on Strengthening the Management of Plastic Packaging Waste in Key Traffic Lines, Watersheds and Tourist Scenic Spots	It is forbidden to use non-degradable disposable foamed plastic tableware in railway stations and passenger trains, passenger ships and tourist ships in inland waters such as the Yangtze River and Tai Lake. Prevent plastic packaging waste and other solid waste from littering and accumulating in rivers, lakes and along the coast. In the Yangtze River, Tai Lake, key tourist attractions (attractions) and other inland waters, the plastic packaging waste that has floated in the water and accumulated on the shore will be cleaned up within three months under the unified leadership of the local people's government in the jurisdiction. It is forbidden to dump garbage along the railway lines, along the Yangtze River and the Tai Lake Basin. All types of ships shall be equipped with sufficient garbage storage containers according to relevant laws, and garbage shall be collected by classification and discharged into garbage reception facilities. Crew members and passengers are prohibited from dumping garbage and cargo residues into rivers (lakes). The competent departments of tourist attractions (attractions) at all levels are responsible for supervising and inspecting the management of plastic packaging waste in the scenic spots (attractions) under their jurisdiction. Each scenic spot (attraction) should be equipped with enough garbage collection containers to facilitate tourists to dispose of garbage. The management unit of tourist attractions (attractions) shall set up special personnel to clean, collect and transport garbage, and maintain garbage collection and storage facilities
	April 2001	Emergency Notice on Immediately Stopping the Production of Disposable Foamed Plastic Tableware	All manufacturing enterprises (including domestic investment, foreign investment and Hong Kong, Macao and Taiwan investment enterprises) should consciously abide by state laws and regulations and implement national industrial policies, and immediately stop producing disposable foamed plastic tableware

continued

Type	Formulated/ Revised Time	Policy Document	Main Content
Normative Documents	January 2002	Notice on Strengthening Law Enforcement and Supervision over the Elimination of Disposable Foamed Plastic Tableware	Local law enforcement departments for industry and commerce, quality inspection and environmental protection shall be conscientiously responsible and, from the date of issuance of this circular, strengthen supervision and inspection over the elimination of disposable foamed plastic tableware in this area in accordance with the law
	December 2007	Notice on Restricting the Production and Sale of Plastic Shopping Bags from the General Office of the State Council	From June 1, 2008, the production, sale and use of plastic shopping bags with a thickness of less than 0.025 mm (hereinafter referred to as ultra-thin plastic shopping bags) are prohibited nationwide. Since June 1, 2008, the system of paid use of plastic shopping bags has been implemented in all commodity retail places such as supermarkets, shopping malls and bazaars, and plastic shopping bags are not allowed to be provided free of charge
	January 2009	Notice on the Control of Excessive Packaging from Commodities of the General Office of the State Council	On the premise of meeting the basic functions of protection, quality assurance, identification and decoration, and in accordance with the principles of reduction, reuse and resource utilization, commodity packaging should be standardized from the aspects of the number of packaging layers, packaging materials, effective volume of packaging, the proportion of packaging costs, and the recycling of packaging materials, so as to guide enterprises to reduce resource consumption in packaging design and production, reduce waste production, and facilitate packaging recycling
	May 2010	Notice on the Construction of Urban Mineral Demonstration from the National Development and Reform Commission and the Ministry of Finance	Through five years of efforts, about 30 "urban mineral" demonstration bases with advanced technology, environmental protection standards, standardized management, large-scale utilization and strong radiation will be built in China. Promote the recycling, large-scale utilization and high-value utilization of key "urban mineral" resources such as scrapped electromechanical equipment, wires and cables, household appliances, automobiles, mobile phones, lead-acid batteries, plastics and rubber
	October 2011	Opinions of the General Office of the State Council on the Establishment of a Complete and Advanced Waste Commodity Recycling System	It is suggested to give full play to the role of the market mechanism to increase the recovery rate of major waste commodities such as scrap metal, waste paper, waste plastic, scrapped automobiles, scrapped mechanical and electrical equipment, scrap tires, scrapped electrical and electronic products, scrapped glass, scrapped lead-acid batteries, and scrapped energy-saving lamps. Strengthen policy guidance and support, further clarify the responsibilities of producers, sellers, and consumers, and effectively recycle key waste commodities through methods such as garbage sorting and recycling

continued

Type	Formulated/Revised Time	Policy Document	Main Content
Normative Documents	January 2013	Notice of the State Council on Printing and Distributing the Circular Economy Development Strategy and Recent Action Plan	It is suggested to implement relevant preferential policies and conduct recycling traditional renewable resources such as scrap metal, waste plastic, waste glass, and waste paper, and increase the recovery rate. The public is encouraged to carry their own shopping bags, and refuse using ultra-thin plastic shopping bags
	April 2013	Notice on Deepening Restrictions on the Use of Plastic Shopping Bags in Production and Sales	The development and reform departments, together with relevant departments, have adopted various forms to vigorously publicize the positive results achieved in saving energy and resources and improving environmental protection awareness since the implementation of the "plastic restriction order" through TV, Internet, radio, newspapers and other media, and advocate green and low-carbon, saving consumption concept. The Education Department should promote the reduction of the use of disposable products such as plastic shopping bags in primary and secondary school students' education planning, and advocate primary and secondary school students to adhere to the concept of conservation and environmental protection in their daily behavior. The Commerce Department, Price Department, and Industry and Commerce Department organized shopping malls, supermarkets, and bazaars to carry out "plastic restriction order" publicity activities, calling on consumers to consciously boycott ultra-thin plastic shopping bags, and promoting operators to consciously implement the paid use system. The business sector organizes relevant associations and enterprises to initiate initiatives to reduce the use of plastic shopping bags. The Ministry of Industry and Information Technology coordinated with telecom operators to send warm reminder text messages to local mobile phone users about the significance of the implementation of the "Plastic Restriction Order" during the fifth anniversary of the implementation of the "Plastic Restriction Order". During the fifth anniversary of the implementation of the "Plastic Restriction Order", the agency affairs management department should organize public institutions to carry out the "Plastic Restriction Order" publicity activities, and advocate the staff of the agency to take the lead in setting an example and reduce the use of plastic shopping bags. The Environmental Protection Department should vigorously publicize the environmental problems caused by ultra-thin plastic shopping bags in conjunction with activities such as "World Environment Day", so that consumers can recognize the harm of ultra-thin plastic shopping bags, and make it a conscious behavior to not use or use plastic shopping bags less. Through platforms such as "Women's Homes" in urban and rural areas, women's federations have adopted colorful and popular methods to publicize the significance of the "Plastic Restriction Order" and the harm of white pollution to the environment and human health, and advocate "use your vegetable basket and you cloth bags"

continued

Type	Formulated/ Revised Time	Policy Document	Main Content
Normative Documents	December 2015	Industry Standard for Comprehensive Utilization of Waste Plastics. Interim Measures for the Administration of the Announcement of Standard Conditions for the Comprehensive Utilization of Waste Plastics	It defines the threshold of waste plastic disposal capacity of the three key types of enterprises newly built and built in the industry
	December 2016	Notice of the General Office of the State Council on Issuing the Implementation Plan of the Extended Producer Responsibility System	Eco-design shall be carried out. Production enterprises should take overall consideration of the resource and environmental impact of raw and auxiliary materials selection, production, packaging, sales, use, recycling, and processing, and carry out in-depth product ecological design. Specifically, it includes lightweight, single, modular, no (low) pollution, easy maintenance design, as well as designs such as life extension, green packaging, energy saving and consumption reduction, and recycling.

Use recycled raw materials. Under the premise of ensuring product quality, performance and safety of use, manufacturers are encouraged to increase the proportion of recycled raw materials used, implement green supply chain management, strengthen the guidance of upstream raw material companies, and develop and popularize technologies for the detection and utilization of recycled raw materials.

Standardize recycling. Production enterprises can standardize the recycling of waste products and packaging through independent recycling, joint recycling or entrusted recycling, and dispose of them directly or by professional enterprises. Responsibility for product recycling and disposal can also be achieved by the production enterprises paying relevant funds in accordance with the law and subsidizing professional enterprises.

Enhance information disclosure. Strengthen the information disclosure responsibility of production enterprises, and make product quality, safety, durability, energy efficiency, toxic and hazardous substance content and other content as mandatory disclosure information to the public; and other contents as targeted disclosure information, which is disclosed to the subjects of waste recycling and resource utilization.

Encourage beverage paper-based composite packaging manufacturers, bottling companies and recycling companies to form alliances in accordance with market-oriented principles, and recycle waste beverages through sales channels of bottling companies, existing renewable resource recovery systems, and recycling companies' self-built networks. Paper-based composite packaging |

continued

Type	Formulated/ Revised Time	Policy Document	Main Content
Normative Documents	December 2016	Guiding Opinions on Accelerating the Development of Renewable Resources Industry	Vigorously promote the construction of waste plastic recycling system, and support the diversified and high-value utilization of different quality waste plastics. Focusing on the current varieties with large resources and high recycling rate, encourage the demonstration of recycling and utilization of key varieties of waste plastics, promote large-scale waste plastics crushing-sorting-modification-granulation advanced and efficient production lines, and cultivate a number of leading enterprises. Actively promote the resource utilization of low-quality and easily polluting waste plastics, encourage the non-polluting energy utilization of domestic waste plastics, and gradually reduce waste plastics to landfill
	April 2017	Circular Development Leads Action	Formulate and publish a list of disposable consumer goods restricted from production and sales and management measures, implement classified management of products included in the directory, and formulate and improve relevant policies for restricting disposable consumer goods. Support the development of reusable alternatives. Research and formulate ecological design standards for disposable products and improve recycling rate
	July 2017	Notice of the General Office of the State Council on Printing and Distributing the Implementation Plan for Prohibiting the Entry of Foreign Waste and Promoting the Reform of the Solid Waste Import Management System	Before the end of 2017, the import of domestic waste plastics, unsorted waste paper, textile waste, vanadium slag and other varieties will be prohibited
	December 2018	Notice of the General Office of the State Council on Printing and Distributing the Pilot Work Plan for the Construction of "Waste-Free Cities"	Recycling, treatment and other parts are the main areas to improve the level of reuse of waste agricultural film and pesticide packaging waste. Establish a recycling system under the guidance of the government, the main body of enterprises, and the participation of farmers. Restrict the production, sale and use of disposable non-degradable plastic bags and plastic tableware, and expand the application scope of degradable plastic products. Accelerate the application of green packaging in the express delivery industry. Promote the use of recyclable items and limit the use of disposable items in hotel, catering and other service industries

continued

Type	Formulated/Revised Time	Policy Document	Main Content
Normative Documents	January 2020	Opinions of the National Development and Reform Commission and the Ministry of Ecology and Environment on Further Strengthening the Control of Plastic Pollution	The production and sale of ultra-thin plastic shopping bags with a thickness of less than 0.025 mm and polyethylene agricultural mulch films with a thickness of less than 0.01 mm are prohibited. The manufacture of plastic products from medical waste is prohibited. The production and sale of disposable foamed plastic tableware and disposable plastic cotton swabs are prohibited; the production of daily chemical products containing plastic microbeads is prohibited. Shopping malls, supermarkets, pharmacies, bookstores and other places, as well as food packaging and takeaway services and various exhibition activities, the use of non-degradable plastic bags is prohibited, and the bazaars regulate and restrict the use of non-degradable plastic bags. The use of non-degradable disposable plastic straws is prohibited in the catering industry. Non-degradable disposable plastic tableware is prohibited for catering services in built-up areas and scenic spots in cities above the prefecture level. Star-rated hotels, regular hotels shall no longer offer disposable plastic products, but can provide related services by setting up self-service purchase machines and providing refillable detergents. Postal express outlets are prohibited from using non-degradable plastic packaging bags, plastic tapes, disposable plastic woven bags, etc. In shopping malls, supermarkets, pharmacies, bookstores and other places, it is recommended to use environmentally friendly cloth bags, paper bags and other non-plastic products and degradable shopping bags. Promote the use of bio-based products such as straw-coated lunch boxes and degradable plastic bags that meet performance and food safety requirements in the field of food delivery. In key mulching areas, the degradable mulch will be promoted on a large scale in combination with agronomic measures. Recyclable and foldable packaging products are encouraged in logistics and distribution industry. With garbage classification regulations in place, the collection and treatment of recyclables such as plastic waste shall be encouraged, and the random stacking and dumping of plastic waste caused by pollution is prohibited. Carry out actions to clean up plastic waste in rivers, lakes and harbors and clean beaches. Promote the cleaning and rectification of farmland residual plastic film, pesticides, fertilizers and plastic packaging, and gradually reduce the amount of farmland residual plastic film
	July 2020	Notice on Actively Promoting Plastic Pollution Control	Published "Refinement Standard for Relevant Plastic Products Prohibition and Restriction Management (2020 edition)"

continued

Type	Formulated/Revised Time	Policy Document	Main Content
Normative Documents	September 2021	"14th Five-Year" Action Plan for Plastic Pollution Control	China will proactively promote the reduction of plastic production and use at the source, including actively promoting the green design of plastic products, continuing to promote the reduction of the use of disposable plastic products, and scientifically and prudently promoting plastic substitute products. China will accelerate the promotion of standardized recycling and disposal of plastic waste, including strengthening the standardized recycling and removal of plastic waste, establishing and improving the collection, transportation and disposal system of rural plastic waste, increasing the recycling of plastic waste, and improving the level of harmless disposal of plastic waste, etc. China will proactively carry out the cleaning and rectification of plastic waste in key areas, and deploy the tasks of cleaning and rectifying plastic waste in rivers, lakes and seas, tourist attractions and rural areas in a targeted manner

Institutes and Enterprises Who Offered Great Support to This Report

China Petroleum and Chemical Industry Federation

China National Resources Recycling Association

Beijing Sankuai Online Technology Co.,Ltd. (Meituan)

KINGFA SCI. & TECH. CO., LTD.

Lenovo (Beijing) Co.

Mars Food (China) Co.

Dow Chemical (China) Co.

ExxonMobil (China) Investment Co.

Acknowledgements

In the process of compiling this report, Bin Xun, Jin Tian, Yang Gao, Xun Gong, Xin Yu, Cong Hou, Chen Bao, Peikun Huang, Zikang Zhao, Yan Wang and more have given great help in data collection, case research, and report compilation. We would like to express the sincere thanks for their enormous support!

At the same time, this report refers to a large number of domestic and foreign literature and reports in the process of preparation, and thanks the researchers for their fruitful work in the treatment of plastic pollution, which provides an important support for the preparation of this research report.